PHYSICS 2
WORKBOOK
for NCEA Level 2

Rob Boasman

Physics 2 Workbook for NCEA Level 2
1st Edition
Rob Boasman

Production controller: Siew Han Ong
Reprint: Natalie Orr
Typeset by Cheryl Rowe, Macarn Design

Any URLs contained in this publication were checked for currency during the production process. Note, however, that the publisher cannot vouch for the ongoing currency of URLs.

Thanks

I am particularly grateful to the following people who have helped me prepare this book: Jenny Thomas and Graham McEwan at Cengage who have demonstrated tremendous patience; Ashwin Reddy and Carol TeBay who spent many hours proof reading and challenged me to do better every step of the way; the teachers and students at Lynfield College who helped me grow as a teacher; my colleagues at Diocesan School for Girls who have been a tremendous support, and without whom I could not have finished this book; Lorraine Rose for keeping my family and I sane during the writing of this book and finally Sarah, Joel, Josh and Sam Boasman for forbearance, support, encouragement and a fresh cup of really hot tea!

Rob Boasman
Diocesan School for Girls
Auckland
2012

For product information and technology assistance,
in Australia call **1300 790 853**;
in New Zealand call **0800 449 725**

For permission to use material from this text or product, please email
aust.permissions@cengage.com

National Library of New Zealand Cataloguing-in-Publication Data
Boasman, Rob.
Physics 2 Workbook / Rob Boasman.

ISBN 978-017019-599-7

1. Physics—Problems, exercises, etc.
I. Title.

530.076—dc 23

Cengage Learning Australia
Level 7, 80 Dorcas Street
South Melbourne, Victoria Australia 3205

Cengage Learning New Zealand
Unit 4B Rosedale Office Park
331 Rosedale Road, Albany, North Shore 0632, NZ

For learning solutions, visit **cengage.co.nz**

Printed in China by 1010 Printing International Limited
16 17 18 25 24

Introduction

The content of the book is based upon the New Zealand Curriculum and the Level 2 NCEA Physics Achievement Standards. The questions in this book have been designed to develop students' ability to apply concepts to increasingly challenging situations, then to practice applying those skills to NCEA style questions. The best starting point for all the exercises is to attempt the worked examples, then compare the solutions to the model answers. Brief answers have been provided for the exercises in the book but fully worked solutions are available on the accompanying Teacher Resource CD.

Throughout this book there are worked examples which have been presented using a strategy known as GUESS, which stands for:

- **Given** – identify the given quantities and convert them into the correct units where necessary.
- **Unknown** – identify the unknown quantity
- **Equation** – identify any equations on the formula page that link the unknown quantity to some or all of the given quantities. Where a single formula does not link all the terms find a second formula that can link to a quantity in the first equation and any unused quantities.
- **Substitute** – substitute the values or changes into the formula and rearrange it where necessary.
- **Solve** – complete all the steps in the equation to solve the problem and consider whether your answer is realistic. The answer should be presented with the relevant unit and to the correct number of significant figures.

The GUESS strategy is effective for both mathematical problems and problems requiring explanation.

CONTENTS

1 Understanding the physical world

1.1 Communicating in science

For students of science to be successful they need more than just a lot of knowledge. They must also develop the skills, attitudes and values to enable them to interact with the wider community and communicate their knowledge in an appropriate and meaningful manner.

Scientific communications use vocabulary, symbols, models, rules and conventions and students must know this scientific language so that they can develop an understanding of socio-scientific issues and express their knowledge and opinions effectively.

Students must be confident in the use of physical quantities, units, Greek symbols and number forms.

Physical quantities and SI units

The study of physics involves the measurement and analysis of **physical quantities**. For consistency of measurement all quantities are measured in S.I. units (an abbreviation of the French 'Système International d'Unités'). The seven fundamental quantities and their units are shown in the table at right.

Quantity		Unit	
Name	Symbol	Name	Symbol
Length	L, d, r	metre	m
Mass	m	kilogram	kg
Time	t	second	s
Temperature	T	kelvin	K
Electric current	I	ampere	A
Luminosity	L	candela	cd
Amount of a substance	n	mole	mol

Derived units

All other physical quantities are derived from the fundamental quantities listed above. The table below shows some of the derived quantities and units that are used in this book.

Derived quantity		Unit		
Name	Symbol	Name	Symbol	Fundamental units
Acceleration	a	metre per second^2		$m\ s^{-2}$
Force	F	newton	N	$kg\ m\ s^{-2}$
Power	P	watt	W	$kg\ m^2\ s^{-3}$
Frequency	f	hertz	Hz	s^{-1}
Potential difference	V	volt	V	$kg\ m^2\ s^{-3}\ A^{-1}$

ISBN: 9780170195997

Symbols

Mathematical symbols and Greek letters are frequently used when writing physics statements or equations. The table below shows some of the symbols that are used and what they mean.

Name	Symbol	Meaning/Use
Equal to...	$=$	Two sides of an equation have the same value
Not equal to...	$\neq$	Two sides of an equation are not the same
Approximately equal to...	$\approx$	Two sides of an equation can be considered to be the same to simplify a solution, e.g. for small angles $\sin \theta \approx \theta$
Exactly equal to...	$\equiv$	Quantities with a precisely defined value, e.g. the mass of a carbon atom is defined as $m_{carbon} \equiv 1.2 \times 10^{-2}$ kg mol^{-1}
Less than...	$<$	The quantity on the left is less than the quantity on the right of an equation
Less than or equal to...	$\leq$	The quantity on the left is less than or equal to the quantity on the right equation
Plus or minus	$\pm$	The quantity could be either positive or negative e.g. $\sqrt{16} = \pm 4$
Change in...	Δ (delta)	Found by calculating the difference between the final and initial values of a changing quantity, e.g. the change in velocity, $\Delta v = v_f - v_i$
Proportional to...	$\propto$	The change in one quantity results in the same sized change in another quantity, e.g. $F_{net} \propto a$ so doubling the net force will double the acceleration
Infinity	∞	a number greater than any real number; without limit
Angles	$\theta \quad \varphi$ (theta) (phi)	Used to indicate angles
Sum of ...	Σ (sigma)	Found by adding all relevant values together e.g. the net force, $F_{net} = \Sigma F = F_1 + F_2 + F_3 + ...$

Number forms

Very large and very small numbers may be expressed in four forms: number, scientific notation, engineering notation or prefix notation.

Form	Notation	Coefficient	Exponent	Example: the speed of light, c
Number	a	a	none	299 792 458
Scientific	$a \times 10^b$	$1 \leq a < 10$	b is any whole number	2.99792458×10^8
Engineering	$a \times 10^b$	$1 \leq a < 1000$	b is any multiple of 3	299.792458×10^6

Very large and very small numbers may also be expressed using prefixes, for example centimetres (cm), where centi means $\times 10^{-2}$. The standard prefixes are shown below.

Prefix	Symbol	Multiplier	Example	Prefix	Symbol	Multiplier	Example
kilo	k	$\times 10^3$	km, kilometres	milli	m	$\times 10^{-3}$	mA, milliamps
mega	M	$\times 10^6$	MHz, megahertz	micro	μ	$\times 10^{-6}$	μT, microtesla
giga	G	$\times 10^9$	GW, gigawatts	nano	n	$\times 10^{-9}$	nm, nanometres
tera	T	$\times 10^{12}$	TB, terabytes	pico	p	$\times 10^{-12}$	pg, picograms

All prefixes must be converted to one of the number forms above, before they are used in equations.

Significant figures

The accuracy of any number is represented by the number of **significant figures** or (s.f.). The rules for applying significant figures are presented below.

Rule	Example	Number of significant figures	Scientific notation
All non-zero digits are significant	43.21	4 s.f.	4.321×10^1
Zeros between numbers are significant	2000.1	5 s.f.	2.0001×10^3
Trailing zeros after a decimal point are significant	0.00900	3 s.f.	9.00×10^3
Leading zeros are **not** significant	0.0007	1 s.f.	7×10^{-4}
Trailing zeros after a number not containing a decimal point *may* or *may not* be significant and must be clarified by an s.f. statement of using scientific notation	500	1 s.f. or 3 s.f.	5×10^2 or 5.00×10^2

When combining numbers by multiplying or dividing, the final answer should be given to the least number of significant figures in the supplied data. However, when combining numbers by adding or subtracting (for example when calculating averages) the final answer should be given to the least number of decimal places.

Mathematics

Physics is a bilingual subject and students must be able to use both language and mathematics skills equally competently. **Appendix 1** contains a number of important mathematical formulas used in Level 2 Physics which students should be familiar with. It also introduces logarithms, a mathematical tool which some students may be familiar with.

Exercise 1A

1   Using the tables on page 4 determine the unit(s) for the following combinations of quantities. Determine the quantity that the combination can be used to calculate.

   a $\dfrac{\text{change in velocity}}{\text{change in time}}$ _____

   b $\dfrac{\text{mass x change in velocity}}{\text{change in time}}$ _____

   c $\dfrac{1}{\text{frequency}}$ _____

   d $\dfrac{\text{power}}{\text{current}}$ _____

2   Convert the following quantities into the fundamental S.I. unit(s) and express them in engineering form.

   a   62 nanoseconds   _____

   b   240 kilovolts   _____

   c   9.9 gigawatts   _____

   d   147 centimetres   _____

   e   583 milligrams   _____

ISBN: 9780170195997

3 Change the following mathematical statements into written statements.

a $a = \dfrac{\Delta v}{\Delta t}$ _____

b $P \propto V$ _____

c $I_{total} = \Sigma I_{individual}$ _____

4 Change the following written statements into mathematical statements.

a The constant, k is defined as being **exactly** 2×10^{-7} T m A^{-1}.

_____

b The forward forces **do not equal** the backwards forces.

_____

c Frequency is **inversely proportional** to time.

_____

5 Round the following numbers to the stated number of significant figures or decimal places.

a 0.000 062 to 1 s.f. _____

b 249 854 to 1 s.f. _____

c 1.009 556 to 4 s.f. _____

6 The following equations have been solved but the answer has not been recorded correctly. In each example write the answer to the correct number of significant figures or decimal places, and explain your decision.

a $452.65 \times 100 = 45\,265$ _____

_____

b $0.2049 \times 1.0 \times 10^{3} = 204.9$ _____

_____

c $15.25 + 16.0 + 15.75 = 47$ _____

_____

7 Speeds are often quoted in km h^{-1} in everyday discussions, but scientists always use m s^{-1}. The ability to convert between the two is important.

a Show that a speed of 1 m s^{-1} is equal to 3.6 km h^{-1}.

_____

_____

_____

_____

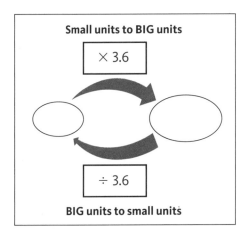

Small units to BIG units

$\times 3.6$

$\div 3.6$

BIG units to small units

ISBN: 9780170195997

b Complete the 'quick converter diagram' on page 7 by adding the units m s^{-1} and km h^{-1} into the correct circle.

Use this relationship to solve the next two problems quickly.

c Convert 50.0 km h^{-1} to m s^{-1}.

_____

d Convert 3.00×10^8 ms^{-1} to km h^{-1}

_____

8 The following questions will require the use of **Appendix 1**. Use Pythagoras's theorem to calculate the length of the missing side of each triangle shown below.

a   **b**   **c**

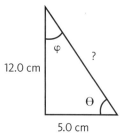 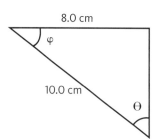 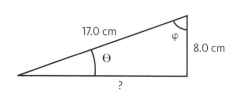

a _____

b _____

c _____

9 For each of the triangles above use trigonometry to calculate the angles Θ and φ.

a _____

b _____

c _____

10 Determine the radius, circumference and area for each of the circles shown below.

        a   **b**   **c**

	a	b	c
Radius, r			
Circumference, C		44 m	
Area, A			50.27 m²

ISBN: 9780170195997

11 Calculate the areas of the triangles and trapezium shown below.

a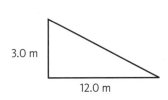

3.0 m

12.0 m

b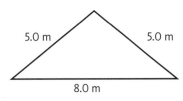

5.0 m   5.0 m

8.0 m

c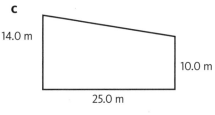

14.0 m

25.0 m

10.0 m

_____   _____   _____

_____   _____   _____

_____   _____   _____

_____   _____   _____

_____   _____   _____

12 Present the following statements in an alternative form by converting the indices, then solve, for example: $5^{-2} = \frac{1}{5^2} = 0.04$.

a $\dfrac{1}{2^5}$ = _____ = _____

b $256^{1/2}$ = _____ = _____

c $729^{-1/3}$ = _____ = _____

d $10^2 \times 10^5$ = _____ = _____

e $10^4 \times 10^8$ = _____ = _____

f $10^6 \times 10^{-4}$ = _____ = _____

g $10^4 \div 10^3$ = _____ = _____

h $10^8 \div 10^9$ = _____ = _____

i $(6^2)^5$ = _____ = _____

j $(2^3)^4$ = _____ = _____

k $(3^4)^{1/2}$ = _____ = _____

13 **Extension:** Present the following statements in an alternative form by converting the logarithm equations, then solve.

a $\log(5) + \log(2)$ = _____ = _____

b $\log(8) + \log(125)$ = _____ = _____

c $\log(18) - \log(6)$ = _____ = _____

d $\log(7) - \log(70)$ = _____ = _____

e $\log(\frac{4}{5})$ = _____ = _____

f $\log(5^2)$ = _____ = _____

g $3\log(3)$ = _____ = _____

ISBN: 9780170195997

2 Experimental techniques

Through experiments and investigations, scientists are able to extend their science knowledge, and develop their understanding of the relationship between investigations and scientific theories and models. Investigative work must be rigorously carried out and presented in such a way that their experiment could be repeated and their findings reviewed.

Planning and gathering data

Introduction

Aim

An aim should inform the reader of the purpose of the experiment or investigation but should not be more than a single sentence long.

Preliminary experiment

Quickly trial the equipment and the experiment to identify relevant physics ideas, maximum and minimum ranges, limitations and control variables.

Hypothesis

Identify any relevant physics ideas and make a prediction.

Method

Independent variable

- State the independent variable (the variable being **changed** by the scientist).
- Identify the range. If the range is reasonable, the graph of the results will show the relationship between the independent and the dependent variables. (As a rough guide, about 70% or more of the measureable range should be used).
- A minimum of five different values of the independent variable should be tested.
- Justify any limitations to the range due to apparatus, the measurements and/or safety.
- Describe and explain any techniques used to increase the accuracy of the measurement of the independent variable (if appropriate).
- Describe any difficulties encountered when measuring the independent variable and discuss how they were overcome.

Dependent variable

- State the dependent variable (the variable being **measured** by the scientist).
- Describe and explain any techniques used to increase the accuracy of the measurement of the dependent variable.
- Describe any difficulties encountered when measuring the dependent variable and discuss how they were overcome.

Control variables

- Describe any other variables that would **significantly** affect the results of the experiment if they are not kept constant, describe how they are controlled and explain why they must be controlled.

ISBN: 9780170195997

Experimental set up

- A fully labelled diagram of the apparatus, showing how it has been set up and how the independent and dependent measurements will be taken.

Experimental techniques

In experimental work the term 'error' is often used to describe the degree of 'uncertainty' in a value due to variations in the measurement that are caused by the lack of precision in a measuring device, or the person using it. The two main types of experimental error are **random** error and **systematic** error.

Random errors

Random errors vary in size and are just as likely to be positive or negative, causing each reading to be spread out around the true value. A good approximation of the true value can be gained by **repeating** the measurement several times and calculating the **average**.

$$\text{average} = \frac{\text{trial 1} + \text{trial 2} + \text{trial 3} + \dots}{\text{number of trials}}$$

Possible random errors include:

Error	Apparatus	Correction	Typical size of error
Reaction time Time taken between an event occurring and the observer reacting.	Stopwatches, clocks, mobile phones.	Repeat the measurement and calculate the average. AND/OR Multiple measurements for periodic events, e.g. a swinging pendulum. $$t_{1\ event} = \frac{t_{multiple\ events}}{\text{number of events}}$$	Estimate to be between 0.1 s and 0.2 s. OR Half the size of the range for each set of repeat readings. $$\text{error} = \frac{t_{max} - t_{min}}{2}$$
Parallax Difficulty in reading analogue meters due to the marker and the scale not being in contact.	Analogue meters, e.g. voltmeters, ammeters, rulers, newton meters.	Ensure that the eye of the observer and the marker and scale are all in line. Some electrical meters have a mirror behind the marker to help with alignment. Repeat the measurement and calculate the average.	Estimate based on the amount the reading changes when the observer moves slightly from side to side.
Counting Occurs when counting multiple events or objects.	The observer. Digital counters.	Repeat the count. OR If the event is random or spontaneous repeat the count and calculate the average.	Eliminated by repeating the count. OR For random or spontaneous event e.g. radioactive decay, use half the size of the range for each set of repeat readings.

Continued over page

ISBN: 9780170195997

Error	Apparatus	Correction	Typical size of error
Division Judging the position of the marker when the value lies between the divisions on the scale.	Analogue meters and digital meters.	Choose a scale that has smaller divisions, e.g. use the mm scale instead of the cm scale on a ruler. AND/OR Multiple measurements for small objects to produce a value more suitable to the scale. AND/OR Repeat the measurement and calculate the average.	Take the error as the smallest division on the scale being used.
Irregularities Density, width, length, etc. is not uniform across the object.	Objects being measured.	Repeat the measurements at several different points, e.g. measure the diameter of a ball across several different diameters and calculate the average.	Half the size of the range for each set of repeat readings.

The significance of the effect of random errors on an experiment can be estimated at the end of an experiment by considering the graph and the proximity (closeness) of the plotted points to the line of best fit. If all the plotted points lie close to the line of best fit then random errors must have been small.

Systematic errors

Most measuring instruments will not be perfect and this will result in errors in the **accuracy** of the readings. A systematic error is one which is constant throughout the experiment, making all the readings too high or all too low. Systematic errors are typically caused by:

Error	Apparatus	Correction
Zero error When the marker doesn't start from zero on the scale.	Analogue meters, e.g. voltmeters, ammeters, rulers, newton meters.	Some electrical meters and newton meters have an adjustment screw which allows the position of the marker needle to be corrected. OR Measure the size of the zero error and add or subtract it (as appropriate) so that the recorded reading is correct.
Calibration error When a device has not been set up to measure correctly.	Analogue meters with incorrectly marked scales, e.g. incorrectly printed meter rulers. Low batteries on digital multimeters. Mobile phone measurement apps.	Initially confirm the readings using an alternative device(s). As it is impossible to determine which device is correct, comparing the similarity will identify any apparatus with a significant calibration error. The device considered most reliable should then be used for all readings.

The size of a systematic error can be estimated at the end of an experiment by comparing the equation of the graph line to a known physics relationship. The difference between the two equations may be due to a systematic error, and provides an opportunity to discuss possible causes.

ISBN: 9780170195997

Working with errors

All errors are quoted to **1 significant figure**, and the decimal place of the error fixes the number of decimal places in the final value. For example: 15.07 ± 0.15 m is correctly written as 15.1 ± 0.2 m

Percentage errors

By comparing the size of the error to the value being measured we can determine if an error is significant or not.

$$\text{percentage error} = \frac{\text{error in reading}}{\text{measured value}} \times 100\%$$

A precise measurement will have an error of 5% or less of the recorded value. A significant error is greater than 10%.

Combining errors

The error in any single measurement will always be the largest error of those contributing to the uncertainty in the value. When two or more measurements are combined, the errors must also be combined to determine the size of the error in the final answer. There are some simple rules to combining errors in Level 2 Physics.

Situation	Rule
Adding or subtracting identical quantities	Add the errors together
Averaging a value	Average the errors
Taking multiple measurements. For example, measuring 1000 sheets of paper and then dividing by 1000 to find the thickness of a single sheet.	Divide the error by the number of multiple events or objects.

Exercise 2A

1   Consider the following measurements and identify possible sources of error that will affect the accuracy and precision of the reading. Suggest techniques that could be used to overcome significant errors.

a   Measuring the thickness of a sheet of paper using a thick wooden metre ruler.

_____

_____

_____

_____

_____

b   Measuring the time period of a pendulum swinging from side-to-side.

_____

_____

_____

_____

_____

c   Measuring the diameter of a ball with a metre ruler.

_____

_____

_____

_____

_____

d   Measuring the time taken for a stone to fall 1.00 m.

_____

_____

_____

_____

_____

2   A small box is measured using a ruler with a scale in millimetres. The readings at either end of the box are 29.5 mm and 42.0 mm. Calculate the length of the box and determine the size of the error in the final answer.

_____

_____

_____

_____

_____

3   A stone is dropped from the top floor of a house and the time taken to reach the ground recorded three times as: 0.99 s, 1.12 s and 0.93 s. Calculate the average time and estimate the size of the error on the readings. Present your final answer to the appropriate number of decimal places.

_____

_____

_____

_____

_____

ISBN: 9780170195997

Data processing

Results tables

Results tables must have sufficient columns to handle all the data, the errors and any processing which is required for multiple readings or averages.

Graphs

Line graphs are always* drawn with the dependent variable on the vertical axis (y-axis) and the independent variable on the horizontal axis (x-axis).

*There are a few exceptions to the rule, for example, time-related quantities are almost always plotted on the horizontal axis regardless of whether the time was the dependent or independent variable.

A graph should include the following:

- Title describing the two variables.
- Axis labelled with quantity and unit.
- Appropriate scale that allows all the points to be plotted accurately, and takes up the majority of the space on the graph paper. Do not use axis breaks (—\/—).
- Points plotted using crosses.
- Anomalous points should be identified by drawing a circle around them.
- A line of best fit (straight or smooth curve).

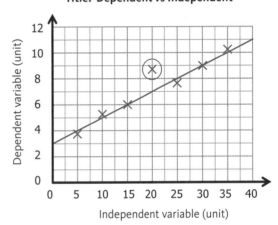

Processing

If the line of best fit is straight then the gradient and intercept can now be calculated and the equation of the line stated.

- The gradient, *m* of a graph is calculated by the equation:

$$m = \frac{\Delta y}{\Delta x} \frac{(y\text{ unit})}{(x\text{ unit})}$$

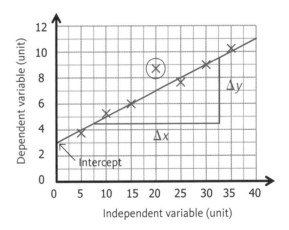

- Draw a triangle on the graph to determine the change in y and the change in x. (The triangle should be more than $^2/_3$ the length of the line of best fit, and touch the line between plotted points).
- The unit of the gradient will be the $\frac{y\text{ unit}}{x\text{ unit}}$.
- The gradient of a straight line is a constant, so is often referred to as the **constant of proportionality**.
- The intercept, *c* is the point at which the line of best fit crosses the y-axis at x = 0. The unit of the intercept will be the unit of the y-axis.
- The equation of a straight line is given by the formula: ***y = mx + c***
- The gradient, intercept and dependent and independent variables must be substituted into the formula along with their units.

ISBN: 9780170195997

Exercise 2B

1 Rob carries out an experiment to determine the effect of increasing current on the size of the magnetic force acting on a 10.0 cm long wire placed in a magnetic field. The measurements are shown below.

a Identify any anomalous values and calculate the averages for the data.

Current (A) (±0.01 A division error)	Force (N) (±0.02 N division error)			
	Trial 1	Trial 2	Trial 3	Average
0.00	0.00	0.00	0.00	
0.30	0.05	0.07	0.07	
0.60	0.15	0.13	0.14	
0.90	0.20	0.18	0.19	
1.20	0.28	0.26	0.32	
1.50	0.33	0.33	0.34	

b Draw a graph of force (y) against current (x) using the data above, and draw the line of best fit.

c State the relationship based upon the shape of the graph.

_____

_____

_____

_____

_____

d Determine the gradient and intercept of the graph and state the equation of the line. Provide units with all your values.

_____

_____

_____

_____

_____

_____

ISBN: 9780170195997

e Compare the equation of the line to the theoretical equation **F = BLI** and hence determine the strength of the magnetic field, B and its unit.

_____

_____

_____

2 Janet is investigating the density (*d*) of glucose. She changes the volume of the fluid in a measuring cylinder which is sitting on a mass balance so that she can measure the mass. The measurements are shown below.

a Identify any anomalous values and calculate the averages for the data.

Volume (cm³) (±0.5 cm³ parallax error)	Mass (g) (±5 g division error)			
	Trial 1	Trial 2	Trial 3	Average
10.0	37	34	36	
20.0	51	47	51	
30.0	62	49	67	
40.0	82	75	77	
50.0	94	89	76	
60.0	109	104	111	
70.0	136	113	120	

b Draw a graph of mass (y) against volume (x) using the data above, and draw the line of best fit.

c State the relationship based upon the shape of the graph.

_____

_____

_____

_____

d Determine the gradient and intercept of the graph and state the equation of the line. Provide units with all your values.

_____

_____

_____

_____

_____

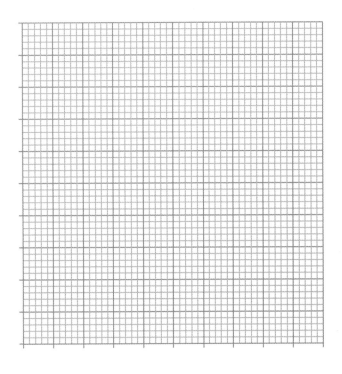

Note: Additional linear graph questions can be found on page 140.

e By considering the method for measuring the mass, suggest why the graph line does not go through the origin (0, 0).

_____

_____

_____

_____

_____

f Compare the equation of the line to the theoretical equation $m = dV$ and hence determine the density (d) of the glucose. Convert your final answer to S.I. units.

_____

_____

_____

_____

_____

Data processing (continued)

Processing non-linear graphs

Many experiments will yield data which does not produce a straight line graph and they cannot be analysed without further processing.

There are four possible curved graph relationships that you will encounter; square, square root, inverse and inverse square. Each one has a characteristic shape, and it is important that you can recognise which relationship is being studied. (The symbol $\propto$ means 'proportional to' and is used to describe a relationship when the exact values are not known.)

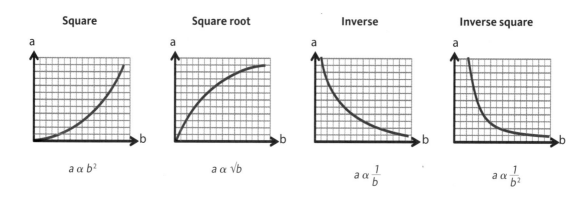

Square	Square root	Inverse	Inverse square
$a \propto b^2$	$a \propto \sqrt{b}$	$a \propto \dfrac{1}{b}$	$a \propto \dfrac{1}{b^2}$

Once the type of relationship has been identified the x axis data can be transformed. For example, for the relationship $a \propto b^2$ all the b values must be squared and a new graph of a against b^2 plotted. The gradient, intercept and equation of the line of best fit can now be determined for the processed graph.

ISBN: 9780170195997

Interpreting

Having completed the experiment and processed the data it is essential that a conclusion is written which covers the following points:

Link the relationship to the aim and physics ideas

- Link the final result with the aim of the experiment, identifying the relationship between the variables.
- Compare the final result with known physics theories to determine the validity of the experiment and whether further work needs to be carried out.

Evaluate the experimental technique (if not previously discussed in the method)

- Discuss any limitations to the experiment or the apparatus used.
- Discuss any controlled variables.
- Explain any accuracy improving techniques.
- Discuss any difficulties that were encountered and how they were overcome.

Evaluate the data

- Discuss inaccurate or anomalous data.
- Discuss the line of best fit, the gradient and the intercept compared to known physics formula.

Exercise 2C

1   Kristina uses a spring to project a glider along an air track. The spring provides a constant force for a constant time. She changes the mass of the glider and measures the speed for each mass using an electronic sensor.

   a   Identify any anomalous values and calculate the averages for the data.

Mass (kg) (±0.002 kg division error)	Speed (ms⁻¹) (±0.5 ms⁻¹ division error)				
	Trial 1	Trial 2	Trial 3	Average	
0.050	23.4	24.3	24.2		
0.100	12.0	13.0	12.2		
0.150	7.8	8.2	8.0		
0.200	5.9	6.1	6.0		
0.250	5.7	4.8	4.9		
0.300	4.0	4.1	4.1		

b Draw a graph of speed (y) against mass (x) using the data above, and draw on the line of best fit.

c State the relationship based upon the shape of the graph.

_____

_____

_____

_____

d Complete the last column of the table by processing the data according to the relationship you have identified. Include an appropriate quantity and unit.

e Plot a second graph using your processed data on the graph below.

f Determine the gradient of the graph and the intercept and state the equation of the line. Provide units with all your values.

_____

_____

_____

_____

_____

_____

_____

g Compare the equation of the line to the theoretical equation $p = mv$ and hence determine the momentum (p) of the glider.

_____

_____

_____

ISBN: 9780170195997

2   Jake spins a small mass around his head. He keeps the radius of the circle constant at 80.0 cm but increases the speed and measures the size of the force using a newton meter. The data is shown below.

a   Identify any anomalous values and calculate the averages for the data.

Speed (ms⁻¹) (±0.5 ms⁻¹ division error)	Force (N) (±0.5 N parallax error)				
	Trial 1	Trial 2	Trial 3	Average	
4.0	1.0	1.0	1.0		
5.0	1.5	1.6	2.5		
6.0	2.2	2.3	2.2		
7.0	3.0	3.2	2.9		
8.0	4.0	4.1	4.0		
9.0	5.0	5.2	4.9		
10.0	7.1	6.3	6.1		

b   Draw a graph of force (y) against speed (x) using the data above, and draw on the line of best fit.

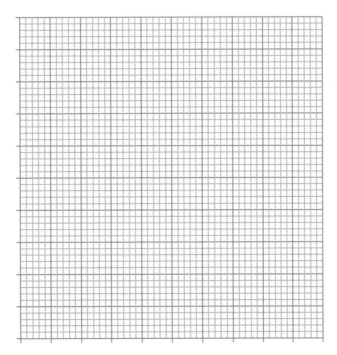

c   State the relationship based upon the shape of the graph.

_____

_____

d   Complete the last column of the table by processing the data according to the relationship you have identified. Include an appropriate quantity and unit.

ISBN: 9780170195997

e Plot a second graph using your processed data on the graph below.

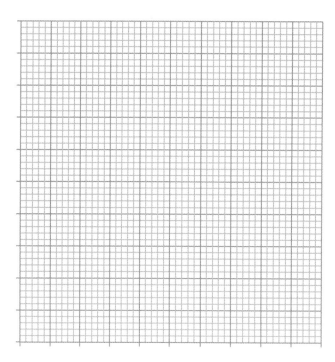

f Determine the gradient of the graph and the intercept and state the equation of the line. Provide units with all your values.

_____

_____

_____

g Compare the equation of the line to the theoretical equation $F = \dfrac{mv^2}{r}$ and hence determine the mass of the object which Jake is spinning around his head.

_____

_____

_____

3 Airini wants to investigate the relationship between the time period of a mass oscillating (bouncing up and down) on a spring and the size of the mass. Her data is shown below.

a Identify any anomalous values and calculate the averages for the 10 oscillations, then determine the average time for a single oscillation.

Mass (kg) (± 0.01 kg division error)	Time for 10 swings (s) (± 0.1 s reaction time error)				Time period (s)	
	Trial 1	Trial 2	Trial 3	Average 10	Average 1	
0.20	4.0	4.1	4.0			
0.40	5.3	5.8	5.8			
0.60	6.9	7.1	7.1			
0.80	8.0	8.2	8.0			
1.00	9.1	9.2	9.0			
1.20	9.8	9.9	10.0			

ISBN: 9780170195997

b Draw a graph of average time for **one oscillation** (y) against mass (x) using the data above, and draw on the line of best fit.

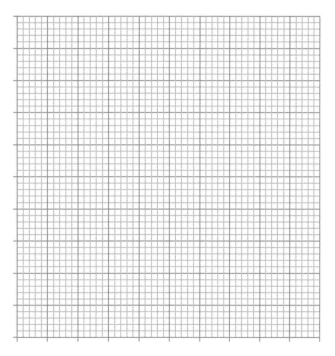

c State the relationship based upon the shape of the graph.

_____

_____

d Complete the last column of the table by processing the data according to the relationship you have identified. Include an appropriate quantity and unit.

e Plot a second graph using your processed data on the graph below.

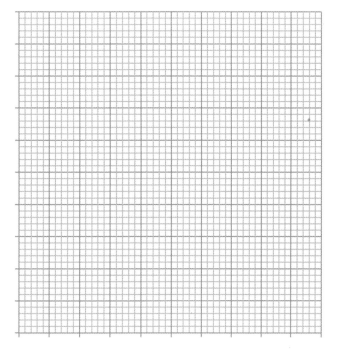

f   Determine the gradient of the graph and the intercept and state the equation of the line. Provide units with all your values.

_____

_____

_____

g   Compare the equation of the line to the theoretical equation $T = \dfrac{2\pi}{\sqrt{k}}\sqrt{m}$ and hence determine the spring constant, k, and its unit.

_____

_____

_____

4   Tama wants to investigate how the strength of the electric field around a small charged ball changes with distance from the ball. The data is shown below.

a   Identify any anomalous values and calculate the averages for the data.

Distance (m) (±0.01 m parallax error)	Electric field strength (Vm⁻¹) (±0.02 Vm⁻¹ division error)			
	Trial 1	Trial 2	Trial 3	Average
0.10	4.38	4.54	4.51	
0.20	1.10	1.15	1.13	
0.30	0.50	0.51	0.50	
0.40	0.28	0.28	0.28	
0.50	0.18	0.18	0.19	
0.60	0.12	0.13	0.13	
0.70	0.09	0.09	0.09	

b   Draw a graph of electric field strength (y) against distance (x) using the data above, and draw on the line of best fit.

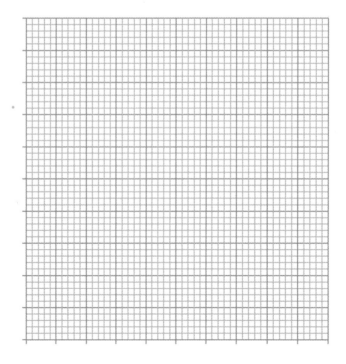

ISBN: 9780170195997

c State the relationship based upon the shape of the graph.

_____

_____

d Complete the last column of the table by processing the data according to the relationship you have identified. Include an appropriate quantity and unit.

e Plot a second graph using your processed data on the graph below.

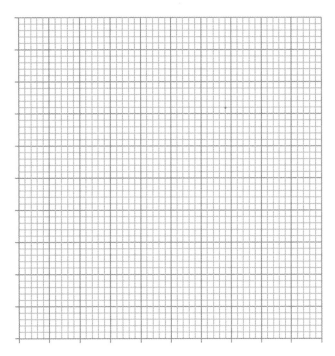

f Determine the gradient of the graph and the intercept and state the equation of the line. Provide units with all your values.

_____

_____

_____

g Compare the equation of the line to the theoretical equation $E = \dfrac{kq}{d^2}$ and given that $k = 8.99 \times 10^9$ V m C^{-1}, calculate the size of the charge on the ball.

_____

_____

_____

Note: Additional non-linear graph questions can be found on pages 62, 78, 122, 147, 195 and 213.

Exercise 2D

Sample reports

The following sample reports contain errors and/or omissions. Read each one and assess them using the mark schedule at the end of each report. You should annotate each report to identify any issues.

Student report 1: Acceleration of a toy car

Aim

To determine the relationship between the distance travelled and the time taken for a toy car moving under the action of a constant force.

Hypothesis

A constant force will cause the toy car to speed up and the equations of motion state that the distance travelled whilst changing speed is given by the formula $d = \frac{(v_i + v_f)}{2} t$.
I predict that the distance will be proportional to the time.

Method

- The independent variable will be the distance travelled along a track. It will range from 20 cm to 140 cm in 30 cm intervals. The maximum distance is limited by the length of the track (140 cm) along which the toy car is rolling.
- The dependent variable will be the time taken to cover each distance which I will measure in seconds using a stopwatch. The time taken to cover each distance will be measured 3 times, and the average calculated.
- During the experiment I will always use the same car and surface so that the friction acting on the car will remain the same.

Diagram of the experimental set-up

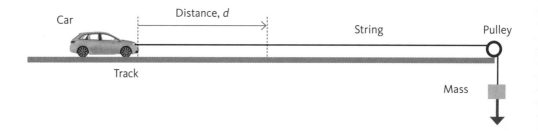

Techniques to improve accuracy

- I will measure the dependent variable three times and calculate the averages. This will help to reduce the effect of random errors due to my reaction time errors and the difficulty of judging exactly when the car has travelled the required distance. My reaction time error will be ± 0.1 s.

ISBN: 9780170195997

Results

The time taken, t, to cover different distances, d, is recorded in the table below.

Distance (cm)	Time (s) (±0.1 s reaction time error)				Time^2 (s^2)
	Trial 1	Trial 2	Trial 3	Average	
20	0.5	0.8	0.6	0.63	0.4
50	0.8	1	1.2	1	1
80	1.4	1.3	1.1	1.26	1.59
110	1.6	1.5	1.4	1.52	2.31
140	1.6	1.8	1.7	1.7	2.89

The averages were calculated and a graph of average distance against time plotted. The shape of the graph reveals the relationship as:

Distance is proportional to time squared.

My prediction was wrong. I will now square all the time values and draw a graph of distance against time squared.

Gradient, m: $m = \dfrac{30}{0.6} = 50 \text{ cm s}^{-2}$

Intercept, c: c = 0.0 cm

Equation of the line:
$$d = 50 \, t^2$$
(cm) (cm s^{-2}) (s^2)

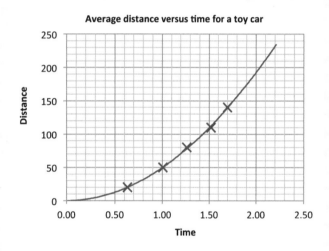

Average distance versus time for a toy car

Conclusion

The experiment went really well as all points lie close to the line of best fit. If I did the experiment again I would choose a shorter maximum length as my mass kept hitting the ground before the car had travelled the full distance. I don't think it had much effect because my graph is so good. The short distances were really hard to measure because they happened so quickly.

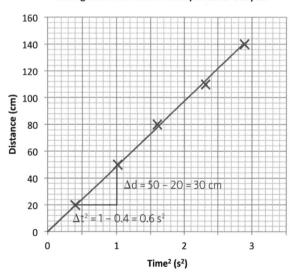

Average distance versus time squared for a toy car

$\Delta d = 50 - 20 = 30 \text{ cm}$

$\Delta t^2 = 1 - 0.4 = 0.6 \text{ s}^2$

Mark sheet for Student report 1

Use the following schedule to mark the student report and determine the appropriate grade.

Identify (circle/underline/annotate) any errors or omissions in the report and write a brief note on the report to explain what is wrong.

	Achieved	✓	Merit	✓	Excellence	✓
Planning and Gathering	• Collect data relevant to the aim based on the manipulation of the independent variable over a reasonable range and number of values: – Distance and time values measured. – Reasonable range (see graph to determine the range is sufficient to show the correct shape). d vs t graph ✓ – 5 or more different values of the independent variable tested.	☐	• Use technique(s) that increase the accuracy of the measured values of the dependent variable: – Repeats and averages (essential). and independent variable, if appropriate: – Parallax error. – Zero error. • Control the variable(s) that could have a significant effect on the results, i.e. – Mass of the car, or; – Mass of the falling weight, or; – Friction (car and surface).	☐ ☐	• Explain why there is a limit to either end of the values chosen for the independent variable. – Short distances are too quick to time accurately. – Long distances limited by the distance the falling mass can travel and/or the length of the ramp – whichever is shorter. • Justify why a variable needs to be controlled. – If the cars mass increases the acceleration will decrease. – If the weight of the falling mass increases the acceleration will increase. – If the friction increases the acceleration will decrease.	☐ ☐
Processing and Interpreting	• Draw a graph that shows the relationship between the independent and dependent variables (see above). – Axis labels and units. – Appropriate linear scales. – Majority of points plotted correctly. – Line of best fit. • Describe the type of mathematical relationship that exists between the variables. – Distance is proportional to time squared.	☐ ☐	• Describe the mathematical relationship obtained from the experimental data. – Gradient, m calculated using the line of best fit (not plotted points). For greater accuracy the gradient points should be near either end of the line of best fit. d vs t^2 ✗ d vs t^2 ✓ *The intercept is not required but useful to identify any systematic errors.* – Equation of the line is of the form: $$d = mt^2 (+ c).$$ Calculated gradient value substituted for m. *(Units are not required but useful to help identify the quantity represented by the gradient.)*	☐	• Discuss any difficulties encountered when making measurements and how these difficulties were overcome. – Small m_{car} and large m_{weight} result in large acceleration so distances are difficult to time accurately – use large m_{car} and small m_{weight}. – Difficult to judge when the car passes the end point – use light gates. • Discuss the relationship between the findings and physics theories/ideas/ formula. – Net force, $F_{net} = F_g - F_{friction}$. – Newton's second law, $F_{net} = ma$. – Equation of motion: $d = \frac{1}{2}at^2$ • Describe any unexpected results and a suggestion of how they could have been caused and/ or the effect they had on the validity of the conclusion.	☐ ☐ ☐
	Achieved: All points correct	☐	**Merit:** All points correct		**Excellence:** 2 good or 3 reasonable points	☐
	Overall Grade					

ISBN: 9780170195997

Student report 2: Acceleration of a toy car

Aim

To determine the relationship between the distance travelled and the time taken for a toy car moving under the action of a constant force.

Hypothesis

A constant force will cause the toy car to accelerate and the equations of motion state that the distance travelled whilst accelerating is given by the formula $d = v_i t + \frac{1}{2}at^2$. As the car is released from rest, the initial velocity will be zero so $d = \frac{1}{2}at^2$. So I predict that $d \sim t^2$.

Method

- The independent variable will be the distance travelled along a track.
  - It will range from 0.60 m to 1.00 m in 0.10 m intervals. I have chosen to use longer distances which take longer times as it makes both the error in the length measurements and the error in timing less significant. The maximum distance is limited by the height of the track which is only 1.20 m above the ground so consequently the falling mass hits the ground and stops pulling the car before the car has travelled the full length of the track.
- The dependent variable will be the time taken to cover each distance.
  - Time will be measured three times using computer light gates and the average time calculated. This will reduce the effect of random errors such as reaction time errors and parallax errors.
- During the experiment there are other variables that must be controlled as they will affect the time taken to cover the distance.
  - Constant mass of the car. As the acceleration of the car is dependent upon its mass it is important to keep the mass of the car constant. A car of mass 0.55 kg will be used throughout the experiment.
  - A small mass will be attached to the car by a long string hanging over a pulley on the end of the track. A mass of 0.100 kg will be used throughout the experiment to provide the constant driving force of $F = mg = 0.100 \times 9.8 = 0.98$ N.

Diagram of the experimental set-up

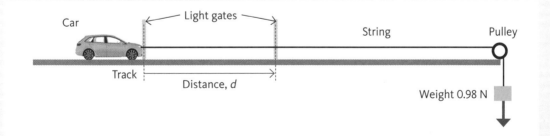

Techniques to improve accuracy

- By taking repeats and averaging I will reduce the effect of random errors, such as releasing the car from the wrong position so that it is already moving when it passes the first light gate, or incorrect placement of the light gates during a trial. The accuracy of the computer light gates will be taken as the smallest division ± 0.01 s.

ISBN: 9780170195997

- To reduce the effect of parallax error when placing the light gates I will ensure that my eye is in line with the gate and the distances marked on the track. Incorrect placement of the light gate could result in the distance travelled being smaller or larger than the recorded distance. My parallax error will be ± 0.01 m.

Results

Distance, (m) (±0.01 m parallax error)	Time (s) (±0.01 s division error)				
	Trial 1	Trial 2	Trial 3	Average	
0.60	2.42	2.49	2.40	2.44	
0.70	2.65	2.68	2.62	2.65	
0.80	2.83	2.85	2.80	2.83	
0.90	2.99	3.01	2.96	2.99	
1.00	3.14	3.20	3.15	3.16	

The averages were calculated and a graph of average distance against time plotted. The shape of the graph reveals the relationship as:

Distance is proportional to time.

This contradicts my prediction that distance is proportional to time squared.

Gradient, m: $\quad m = \dfrac{0.72}{1.3} = 0.55 \text{ m s}^{-1}$

Intercept, c: $\quad c = -0.76$ m

Equation of the line:
$$d = 0.55\, t - 0.76$$
$$\text{(m)} \quad \text{(m s}^{-1}\text{) (s)} \quad \text{(m)}$$

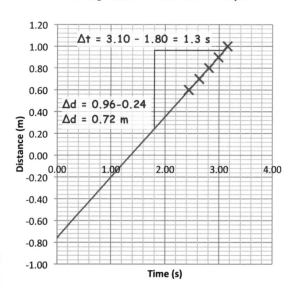

Average distance versus time for a toy car

Conclusion

The data is very precise as all points lie close to the line of best fit, however, my results and graph suggest that the car is moving at a steady speed, which I know it didn't because the car was released from rest and clearly accelerated. The forces on the car may have become balanced really quickly, which would explain why the car was travelling at a steady speed when it reached a distance of 0.60 cm. If I were to repeat the experiment I would use a bigger range of independent values to see if it revealed the shape of the relationship more clearly so that I can identify when the car is accelerating.

ISBN: 9780170195997

Mark sheet for Student report 2

Use the following schedule to mark the student report and determine the appropriate grade.

Identify (circle/underline/annotate) any errors or omissions in the report and write a brief note on the report to explain what is wrong.

	Achieved	✓	Merit	✓	Excellence	✓
Planning and Gathering	• Collect data relevant to the aim based on the manipulation of the independent variable over a reasonable range and number of values: – Distance and time values measured. – Reasonable range (see graph to determine the range is sufficient to show the correct shape). *[graph d vs t ✓]* – 5 or more different values of the independent variable tested.	☐	• Use technique(s) that increase the accuracy of the measured values of the dependent variable: – Repeats and averages (essential). and independent variable, if appropriate: – Parallax error. – Zero error. • Control the variable(s) that could have a significant effect on the results, i.e. – Mass of the car, or; – Mass of the falling weight, or; – Friction (car and surface).	☐ ☐	• Explain why there is a limit to either end of the values chosen for the independent variable. – Short distances are too quick to time accurately. – Long distances limited by the distance the falling mass can travel and/or the length of the ramp – whichever is shorter. • Justify why a variable needs to be controlled. – If the cars mass increases the acceleration will decrease. – If the weight of the falling mass increases the acceleration will increase. – If the friction increases the acceleration will decrease.	☐ ☐
Processing and Interpreting	• Draw a graph that shows the relationship between the independent and dependent variables (see above). – Axis labels and units. – Appropriate linear scales. – Majority of points plotted correctly. – Line of best fit. • Describe the type of mathematical relationship that exists between the variables. – Distance is proportional to time squared.	☐ ☐	• Describe the mathematical relationship obtained from the experimental data. – Gradient, m calculated using the line of best fit (not plotted points). For greater accuracy the gradient points should be near either end of the line of best fit. *[graph d vs t^2 ✗] [graph d vs t^2 ✓]* *The intercept is not required but useful to identify any systematic errors.* – Equation of the line is of the form: $$d = mt^2 (+ c).$$ Calculated gradient value substituted for m. *(Units are not required but useful to help identify the quantity represented by the gradient.)*	☐	• Discuss any difficulties encountered when making measurements and how these difficulties were overcome. – Small m_{car} and large m_{weight} result in large acceleration so distances are difficult to time accurately – use large m_{car} and small m_{weight}. – Difficult to judge when the car passes the end point – use light gates. • Discuss the relationship between the findings and physics theories/ideas/ formula. – Net force, $F_{net} = F_g - F_{friction}$. – Newton's second law, $F_{net} = ma$. – Equation of motion: $d = \frac{1}{2}at^2$ • Describe any unexpected results and a suggestion of how they could have been caused and/or the effect they had on the validity of the conclusion.	☐ ☐ ☐
	Achieved: All points correct		**Merit:** All points correct		**Excellence:** 2 good or 3 reasonable points	
	Overall Grade					

ISBN: 9780170195997

Further questions about Student report 2

1   After writing the report, Student 2 made additional measurements which are shown below.

   a   Identify any anomalous values and calculate the averages for the new data.

Distance (m) (±0.01 m parallax error)	Time (s) (±0.01 s division error)				
	Trial 1	Trial 2	Trial 3	Average	
0.10	0.98	1.00	0.98		
0.20	1.24	1.42	1.38		
0.30	1.70	1.72	1.74		
0.40	1.98	2.00	1.79		
0.50	2.25	2.22	2.22		
0.60	2.42	2.49	2.40	2.44	
0.70	2.65	2.68	2.62	2.65	
0.80	2.83	2.85	2.80	2.83	
0.90	2.99	3.01	2.96	2.99	
1.00	3.14	3.20	3.15	3.16	

   b   Draw a graph of distance against time using the complete set of data above, and draw on the line of best fit.

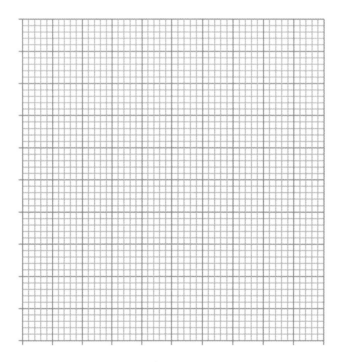

   c   State the relationship based upon the shape of the graph.

   _____

   _____

   d   Complete the table by processing the data to produce a linear relationship.

ISBN: 9780170195997

e   Plot a graph of the processed data.

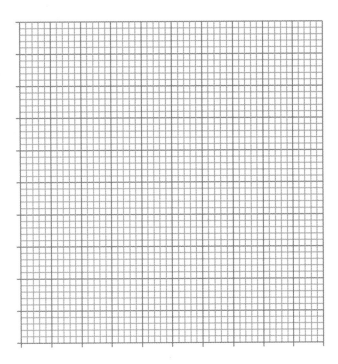

f   Determine the gradient of the graph and the intercept and state the equation of the line.
Provide units with all your values.

_____

_____

_____

_____

g   Compare the equation of the line to Student 2's hypothesis, which suggested that $d = \frac{1}{2}at^2$
and hence determine the acceleration of the car along the track.

_____

_____

_____

h   Calculate the net force on the whole system (the car and the falling mass) using the
acceleration you calculated in **g**.

_____

_____

i   By considering the size of the force due to the falling mass and the net force on the whole
system determine the size of the friction force acting on the system.

_____

_____

ISBN: 9780170195997

3 Light and waves

3.1 Properties of an image

The image produced by an optical instrument can be described in terms of its **magnification**, **orientation** and **nature**.

Term		Definition
Magnification or size (*m*)	Diminished	The image is smaller than the original object
	Same size	The image is same size as the object
	Magnified	The image is larger than the object
Orientation	Upright	The image is same way up as the object
	Inverted	The image is upside down compared to the object
Nature	Real	Rays really pass through the image
	Virtual	Rays only appear to come from the image

3.2 The laws of reflection

When any ray of light strikes a surface and is reflected, the direction of the reflected ray can be determined by using the two laws of reflection:

Law 1 The angle of incidence (*i*) EQUALS the angle of reflection (*r*).

Law 2 The incident ray, the reflected ray and the normal all lie on the same plane.

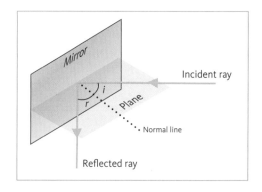

3.3 Spherical mirrors

Concave mirrors (converging)
Rays parallel and close to the principal axis will, upon reflection, converge on the principal focus (F) of the mirror.

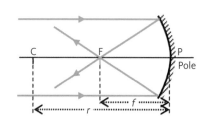

P Optical centre of the mirror
F Principal focus (**real**)
f Focal length (**positive**)
C Centre of curvature (**real**)
r Radius of curvature (**positive**)

$$r = 2f$$

ISBN: 9780170195997

Convex mirrors (diverging)
Rays parallel and close to the principal axis will, upon reflection, form a diverging beam that appears to come from the principal focus (F) of the mirror.

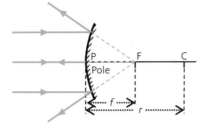

P	Optical centre of the mirror
F	Principal focus (**virtual**)
f	Focal length (**negative**)
C	Centre of curvature (**virtual**)
r	Radius of curvature (**negative**)

$$r = 2f$$

Ray diagrams

The position, size and nature of the **image** (I) of an **object** (O) produced by a spherical mirror can be found by drawing a ray diagram using four rules:

Rule 1 A ray parallel to the principal axis is reflected through the principal focus (or appears to diverge from it).

Rule 2 A ray through (or from) the principal focus is reflected parallel to the principal axis.

Rule 3 A ray incident on the pole of the mirror is reflected at an angle equal to the angle of incidence.

Rule 4 A ray through (or from) the centre of curvature is reflected back along its own path.

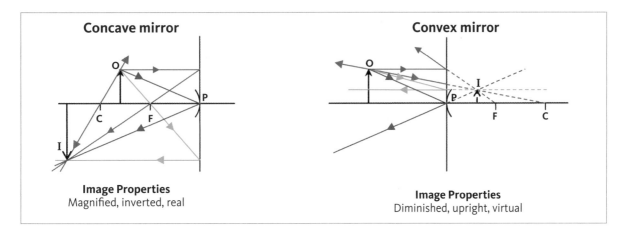

Concave mirror

Image Properties
Magnified, inverted, real

Convex mirror

Image Properties
Diminished, upright, virtual

Any two rays are sufficient to determine the position of the image. A third ray can be drawn to confirm the answer.

Mirror formula

The object and its image are related mathematically in terms of their position and size, and the focal length of the mirror.

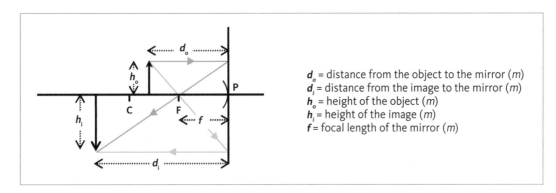

d_o = distance from the object to the mirror (*m*)
d_i = distance from the image to the mirror (*m*)
h_o = height of the object (*m*)
h_i = height of the image (*m*)
f = focal length of the mirror (*m*)

ISBN: 9780170195997

Magnification formula

To determine the magnification (*m*) produced by a mirror:
$$m = \frac{d_i}{d_o} = \frac{h_i}{h_o}$$

Focal length formula

To determine the position of an image, an object or the focal length of a mirror:
$$\frac{1}{f} = \frac{1}{d_i} + \frac{1}{d_o}$$

Real-is-positive convention

When solving mirror formula problems it is essential to substitute the appropriate sign(s) into the equations. The value and sign of the answer will then provide information about its position and nature.

- Concave mirrors have a **real** principal focus so the focal length (*f*) is **positive**.
- **Real** objects and images have **positive** distances.
- Convex mirrors have a **virtual** principal focus so the focal length (*f*) is **negative**.
- **Virtual** objects and images have **negative** distances.

Magnifications (*m*) are always quoted as **positive** in questions, but the magnification of a virtual image should be substituted into equations as a **negative** value.

Worked example: Mirror ray diagram 1

Determine the position and properties of the image formed by a 1.0 cm tall object placed 5.0 cm from a concave mirror of focal length 3.0 cm.

Solution

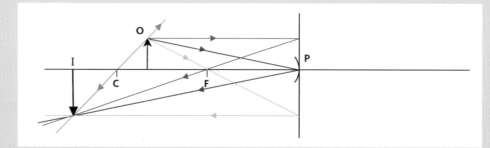

*A ray diagram must first be constructed for a **concave** mirror with P, F and C clearly marked. From the question:*

$f = + 3.0$ cm   *(concave mirrors have **positive** focal lengths)*
so $r = + 6.0$ cm   *(as r = 2f and is also **positive**)*

The object should then be added to the diagram. It is helpful to apply the same scale to both distance and height, though not essential.

$d_o = + 5.0$ cm   *(object is real so has a **positive** distance)*
$h_o = 1.0$ cm

By applying all four rules it is possible to determine the position of the image, from which the properties can be determined.
   As all rays pass through the image it is described as real, and a solid arrow added to the diagram to show its position.

ISBN: 9780170195997

By measurement: Magnification $= \dfrac{h_i}{h_o} = \dfrac{1.2 \text{ cm}}{0.8 \text{ cm}} = 1.5$

Properties: The image is magnified, inverted and real.

Exercise 3A

1   Complete the following ray diagrams by applying all four rules. (Scale: One square = 0.5 cm)

   a   Find the position and properties of the image formed by a 2.0 cm tall object placed 7.0 cm from a **concave** mirror of focal length 2.5 cm.

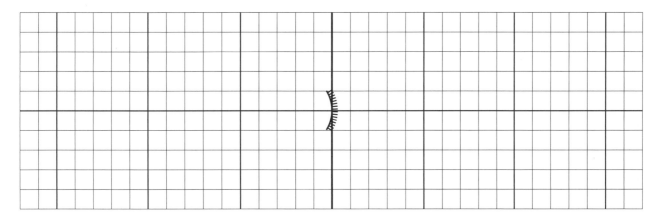

Properties of the image:

Nature: Size: Orientation: Magnification:

b Find the position and properties of the image formed by a 2.0 cm tall object placed 7.0 cm from a **convex** mirror of focal length 2.5 cm.

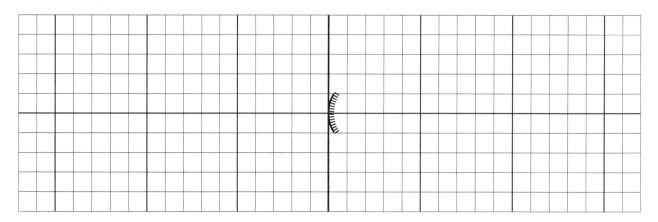

Properties of the image:

Nature: Size: Orientation: Magnification:

c Find the position and properties of the image formed by a 1.0 cm tall object placed 1.0 cm from a **concave** mirror of focal length 5.0 cm.

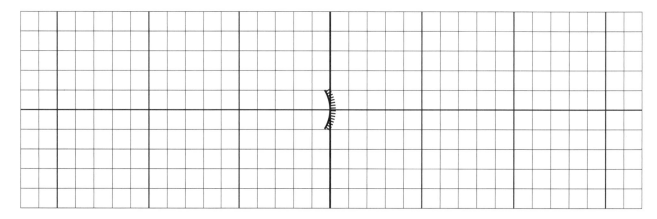

Properties of the image:

Nature: Size: Orientation: Magnification:

ISBN: 9780170195997

d   The object and image are shown on the diagram below. State the nature of the image and then using ray diagrams, determine the focal length and radius of curvature of the **concave** mirror.

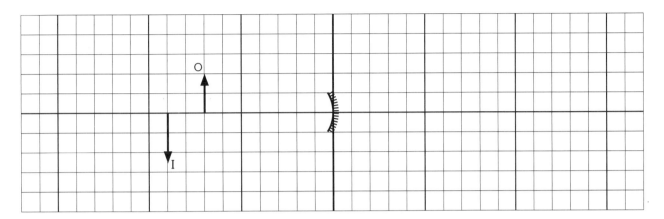

Properties of the image:

Nature: Focal length: Radius of curvature:

e   The object and image are shown on the diagram below. State the nature of the image and then using the ray rules, determine the focal length and radius of curvature of the **convex** mirror.

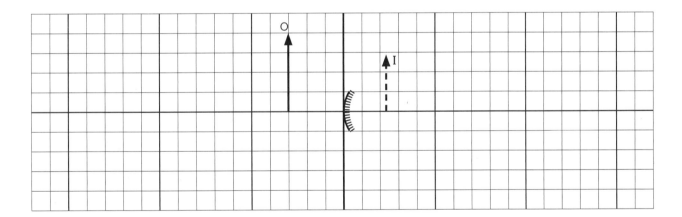

Properties of the image:

Nature: Focal length: Radius of curvature:

ISBN: 9780170195997

f Find the size and position of an object which forms a 2.0 cm tall virtual image in a concave mirror. The image distance is the same size as the focal length (f) = 4.0 cm.

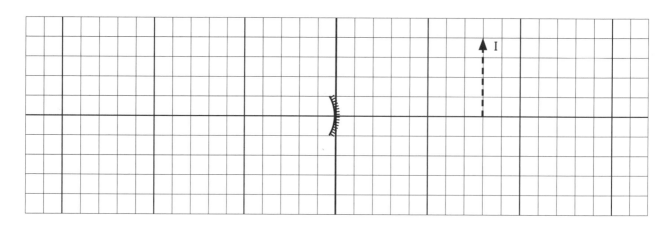

Properties of the object:

Size: Object distance:

2 Use the mirror formula $\dfrac{1}{f} = \dfrac{1}{d_i} + \dfrac{1}{d_o}$ to solve the following problems.

a Geetasha places a concave mirror 60 cm away from a candle and focuses a real image onto a piece of card 40 cm away from the mirror. Determine the focal length of the mirror.

_____

_____

_____

_____

b Matthew is standing 4.8 m away from a shop security mirror. The mirror is convex and has a focal length of 7.2 m. Calculate the distance of Matthew's virtual image from the pole of the mirror.

_____

_____

_____

_____

c Using a concave mirror of focal length 5.0 m Ashwin is able to produce an image of a distant tower on a screen 5.1 m away from the pole of the mirror. Calculate the distance of the tower from the mirror.

_____

_____

_____

_____

ISBN: 9780170195997

3   Use $\dfrac{1}{f} = \dfrac{1}{d_i} + \dfrac{1}{d_o}$ and $m = \dfrac{d_i}{d_o} = \dfrac{h_i}{h_o}$ to solve the following problems.

   a   A concave mirror of focal length 18 cm is used to produce a real image of an object placed 42 cm from the pole of the mirror. Given that the object is 9.0 cm tall, determine:

      i   the distance of the image from the mirror

      ii   the magnification

      iii   the size of the image.

   b   Andrew can see a 6.4 m tall tree from his bedroom 13 m away. He uses a concave mirror to form a real image of the tree on a screen. The image is 42 cm tall. Calculate:

      i   the magnification of the image

      ii   the distance of the image from the pole of the mirror

      iii   the focal length of the mirror.

4   Use $r = 2f$, $\dfrac{1}{f} = \dfrac{1}{d_i} + \dfrac{1}{d_o}$ and $m = \dfrac{d_i}{d_o} = \dfrac{h_i}{h_o}$ to solve the following problems.

   a   Divya is 0.84 m tall and stands 1.7 m away from a convex mirror with a radius of curvature of 10.2 m. Calculate:

      i   the focal length

      ii   the distance of Divya's image from the mirror

      iii   the magnification

      iv   the height of the image.

ISBN: 9780170195997

b Katie can see her image in a convex mirror of radius of curvature 6.6 m. Her image is 41 cm tall and appears to be 2.4 m inside the mirror. Calculate:
  i the distance of Katie from the pole of the mirror
  ii the magnification of the image
  iii Katie's real height.

_____

_____

_____

_____

_____

_____

5 Philip uses a convex mirror of focal length 32 cm to produce an image of a book placed at a distance, d_o in front of the mirror. Given that the image is half the size of the book, calculate the distance between the book and its image.

_____

_____

_____

_____

_____

_____

_____

_____

6 Sue uses a concave mirror with a focal length of 15 cm to produce an image of a flower that is twice the size of the actual flower. Determine the two possible positions where Sue could place the flower to achieve these images.

_____

_____

_____

_____

_____

_____

_____

ISBN: 9780170195997

7   Tony is studying reflection. He places a mirror opposite to a wall that has been painted white. Tony can see a reflection of the ceiling lamp in the mirror but not in the wall. Complete the following diagram for rays striking the two plane surfaces, and use your diagram to explain why a clear image can be seen in a mirror but not in a wall. For each surface you should draw:

*   Normals where the rays strike the surface.
*   Reflected rays that obey the laws of reflection.
*   Construction lines inside the surface to show where the ray appears to have come from.

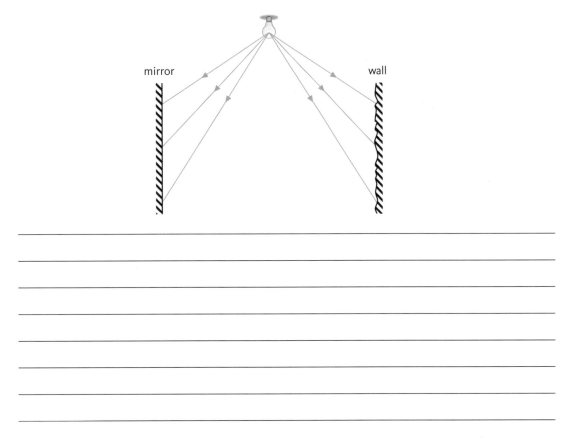

_____

_____

_____

_____

_____

_____

_____

_____

_____

8   With the aid of two diagrams, explain how you could find the principal focus of a concave mirror and a convex mirror. Your diagrams should be clearly labelled with the pole (P), principal focus (F), centre of curvature (C), focal length (f) and the radius of curvature (r.)

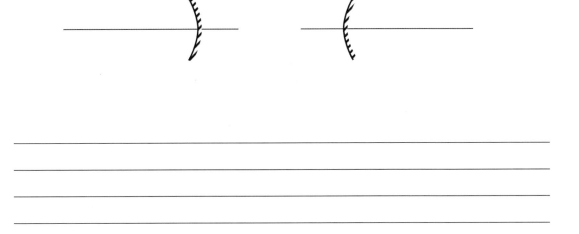

_____

_____

_____

_____

ISBN: 9780170195997

9 The wing mirror on Jigna's car is broken, so she tries to fix it using a plane mirror. Using the laws of reflection, complete the following diagrams to determine the difference in ray paths for a plane mirror and a convex mirror. Discuss the images that Jigna sees in each mirror. Your discussion should include the properties of the image and compare the size of the field of view each mirror produces.

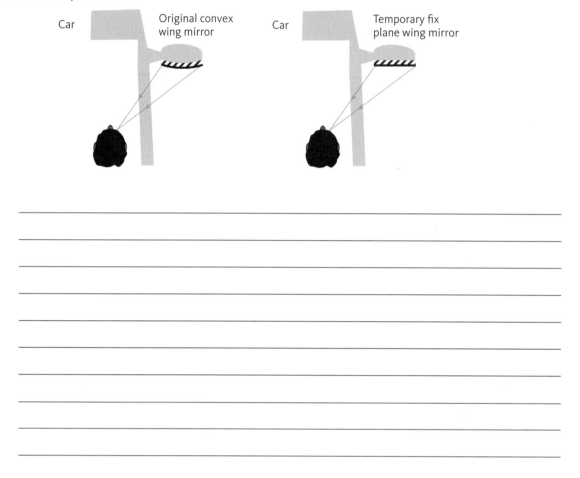

_____

_____

_____

_____

_____

_____

_____

_____

_____

_____

_____

3.4 Refraction

When light travels from a low optically dense medium, such as air, into a high optically dense medium, such as glass, it slows down. The greater the **increase** in optical density, the greater the **decrease** in the speed of the light. This is described as an inversely proportional relationship.

$$\frac{\text{refractive index for medium 1}}{\text{refractive index for medium 2}} = \frac{\text{speed of light in medium 2 (m s}^{-1}\text{)}}{\text{speed of light in medium 1 (m s}^{-1}\text{)}}$$

$$\frac{n_1}{n_2} = \frac{v_2}{v_1}$$

If the incident ray strikes the boundary at an angle, then this change in speed also causes the ray to change direction. The light is described as being **refracted**.

ISBN: 9780170195997

The laws of refraction

Law 1   For a particular colour of light, the ratio of the sine of the angle is inversely proportional to the ratio of the refractive index as the ray travels from one medium to the other, so we have:

$$n_1 \sin \theta_1 = n_2 \sin \theta_2$$

This is also known as Snell's Law.

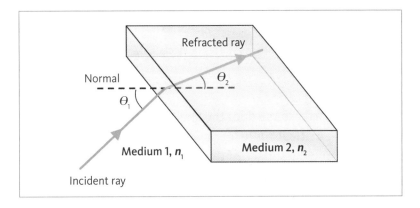

Law 2   The incident ray, the normal and the refracted ray all lie on the same plane.

Combining Snell's Law and the relationship between velocity and refractive index gives us:

$$\frac{n_1}{n_2} = \frac{v_2}{v_1} = \frac{\sin \theta_2}{\sin \theta_1}$$

Dispersion

The refractive index of a material depends on the colour (frequency) of the light incident upon it. This means that different colours are refracted by different amounts.

   This can be clearly seen when white light is incident upon an equilateral prism at an angle. The white light is observed to split up or disperse into a band of colours known as a **spectrum**.

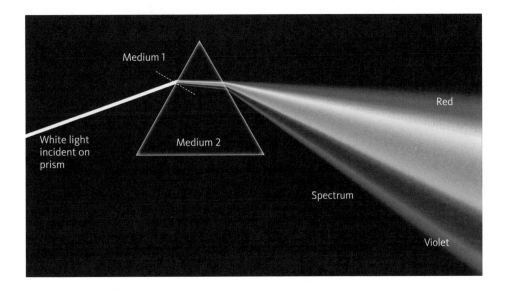

ISBN: 9780170195997

Total internal reflection and the critical angle

When light travels from a **high optically dense to a low optically dense** medium, the ray is refracted away from the normal and a weak ray is reflected internally back into the denser medium (Incident ray 1).

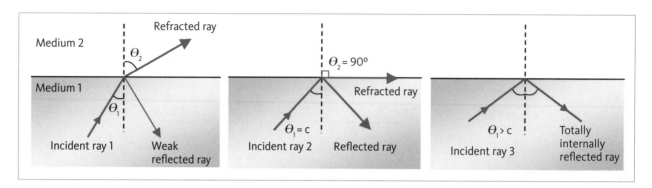

As the angle of incidence (θ_1) increases so does the angle of refraction (θ_2), until a certain angle of incidence is reached, called the **critical angle** (c), at which point the angle of refraction $\theta_2 = 90°$, and a weak refracted ray passes along the surface of the denser medium (Incident ray 2).

The critical angle can be calculated by comparing the refractive index of the two media.

$$\sin \theta_c = \frac{n_2}{n_1}$$

For angles of incidence greater than this critical angle (c), the refracted ray disappears and all the light is **totally internally reflected** (TIR) (Incident ray 3), resulting in a bright reflected ray.

Worked example: The laws of refraction

A ray from a swimming pool light is incident upon the surface of the water at an angle of 65° to the surface, as shown in the diagram.
Calculate the angle of refraction of the ray in the air.

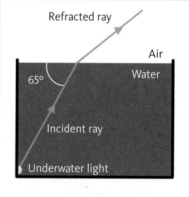

Absolute refractive index of air, $n_a = 1.00$
Absolute refractive index of water, $n_w = 1.33$

Solution

Watch out! Angles must always be measured to the normal. Whenever an angle is given that is measured to the surface we must first determine the incident angle as measured to the normal.

ISBN: 9780170195997

Given n_a = 1.00

          n_w = 1.33

          angle to the surface = 65°

Unknown Θ_a = ?

Equations $n_1 \sin \Theta_1$ = $n_2 \sin \Theta_2$

Substitute Θ_w must be measured to the normal

          Θ_w = 90 − 65 = 25°

          $n_w \sin \Theta_w$ = $n_a \sin \Theta_a$

          1.33 sin 25 = 1.00 sin Θ_a

          $\sin \Theta_a$ = $\dfrac{1.33 \sin 25}{1.00}$ = 0.562

          Θ_a = $\sin^{-1}(0.562)$

Solve Θ_a = **34° (2 s.f. due to the angle)**

*Consider your answer: When the ray enters the air it will speed up, and since the angle of refraction is **proportional** to the velocity we would expect the angle to get bigger. This agrees with our solution.*

Exercise 3B

1   Complete the following ray paths for a ray of yellow light travelling through each block. You should draw a normal at the point where the ray is incident upon each boundary, and include both refracted rays and internally reflected rays.

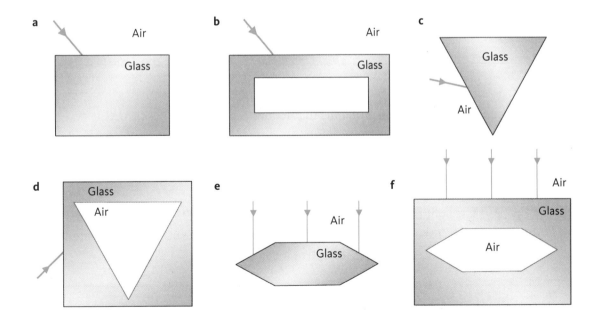

ISBN: 9780170195997

2 By considering these equations, describe what will happen in the situations that follow, stating which equation informed your decision:

$$\frac{n_1}{n_2} = \frac{\sin\Theta_2}{\sin\Theta_1} \qquad \frac{n_1}{n_2} = \frac{v_2}{v_1} \qquad \frac{\sin\Theta_1}{\sin\Theta_2} = \frac{v_1}{v_2}$$

a A ray is incident upon a more optically dense medium normal to the surface ($\Theta_i = 0°$.) Describe the change in velocity and angle to the normal in the new medium.

_____

_____

_____

b A ray is incident upon a more optically dense medium at an angle to the surface. Describe the change in velocity and angle to the normal in the new medium.

_____

_____

_____

c A ray enters a new medium at an angle and speeds up. Describe the change in refractive index and angle to the normal.

_____

_____

_____

d A ray enters a new medium and the angle to the normal decreases. Describe the change in velocity and refractive index.

_____

_____

_____

3 Use these equations to solve the problems that follow.

$$\frac{n_1}{n_2} = \frac{\sin\Theta_2}{\sin\Theta_1} \qquad \frac{n_1}{n_2} = \frac{v_2}{v_1} \qquad \frac{\sin\Theta_1}{\sin\Theta_2} = \frac{v_1}{v_2}$$

a A ray of yellow light is travelling through water at a speed of 2.25×10^8 m s^{-1} when it is incident upon a sheet of ice, causing it to speed up to 2.29×10^8 m s^{-1}. Calculate the refractive index of ice. (Absolute refractive index of water, $n_w = 1.33$.)

_____

_____

_____

ISBN: 9780170195997

b A ray of yellow light in air is incident upon a glass block at an angle of 57.8° to the normal and is refracted at an angle of 31.9°. Calculate the absolute refractive index of the glass. (Absolute refractive index of air, n_a = 1.0.)

_____

_____

_____

_____

c A ray of yellow light **in water** is incident upon the surface at an angle of 22.0° to the normal and is refracted at an angle of 29.9° to the normal. Calculate the speed of yellow light in the water. (Speed of light in air, v_a = 3.0 × 10⁸ m s⁻¹.)

_____

_____

_____

_____

d Crystalline silicon has a refractive index of n_s = 3.97 for yellow light. Determine the change in the speed of light if a ray of yellow light travels from air into a sheet of crystalline silicon. (Absolute refractive index of air, n_a = 1.0008; velocity of light in air, v_a = 2.996 × 10⁸ m s⁻¹.)

_____

_____

_____

_____

_____

_____

e A ray of yellow light travels from inside an acrylic glass block into air. It is incident upon the boundary at an angle of 28° to the normal. Calculate the angle of refraction and the angle through which the ray is deviated. (Absolute refractive index of air, n_a = 1.00; absolute refractive index of acrylic glass, n_g = 1.49.)

_____

_____

_____

_____

_____

_____

_____

4 In Borneo an orang-utan was photographed attempting to catch fish using a spear – behaviour it had learned from observing local fishermen. But even orang-utans need to understand physics if they want to catch fish! On the diagram below are three rays of light travelling from the nose of the fish to the surface. The rays strike the surface at 30°, 40° and 50° to the normals (n_a = 1.00, n_w = 1.33).

a Using Snell's Law, determine the path of the light rays after they are incident upon the surface.

_____

_____

_____

_____

_____

_____

_____

_____

b Draw the paths onto the diagram and use them to find the position of the image of the fish.

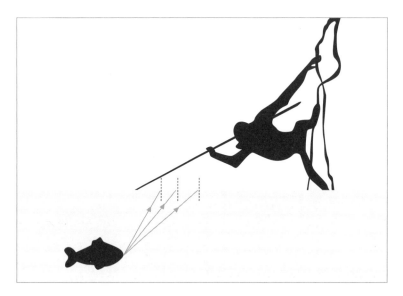

c Determine if the orang-utan will get his fish, and if not suggest how he could increase his chance of success?

_____

_____

_____

_____

_____

ISBN: 9780170195997

5 Safety glass is constructed from three layers made up of two identical layers of glass with a layer of polyvinyl butyral (PVB) binding the two sheets of glass together. (Absolute refractive index of air, n_a = 1.00; absolute refractive index of glass, n_g = 1.55; absolute refractive index of PVB, n_p = 1.48; speed of light in air = 3.00 × 10^8 m s^{-1}.)

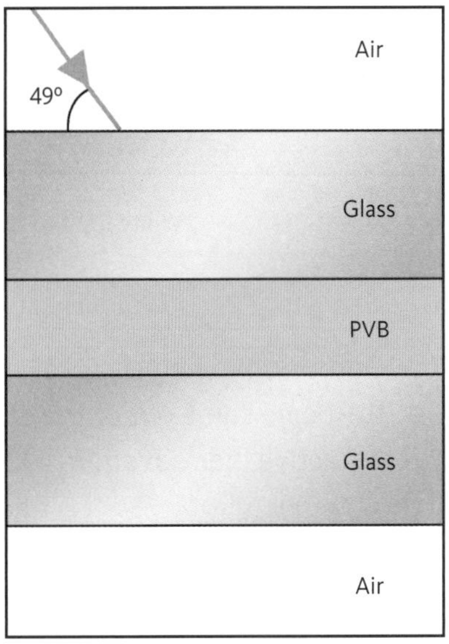

a Complete the diagram at right to show the path of a ray through the safety glass.

b Calculate the speed of light in the PVB.

_____

_____

_____

_____

_____

_____

c Calculate the angle of refraction in the lower glass layer.

_____

_____

_____

_____

6 When a ray of **white light** passes through a prism it splits up to form a rainbow.

a Name the effect that causes white light to form a spectrum, and explain why it occurs.

_____

_____

_____

_____

_____

_____

_____

_____

_____

_____

_____

b Complete the diagram to show what would happen to the ray of white light that is incident upon an equilateral prism.

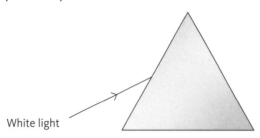

White light

7 Gordon shines a yellow ray into a solid glass fibre as shown in the diagram below, and observes the beam come out of the other end. (Absolute refractive index of glass, n_g = 1.48; absolute refractive index of air, n_a = 1.00.)

Glass

Air

a Draw the path of the ray until it leaves the glass fibre.

b Gordon tries to make his own optic fibre using a glass tube, as shown in the diagram below. Explain why Gordon's glass tube will not work as an optic fibre and draw the path of the ray until it leaves the glass tube.

   •   Your explanation should state the physics ideas behind the optic fibre and include the conditions necessary for this to occur.

Glass

Air

Glass

_____

_____

_____

_____

_____

_____

_____

_____

_____

_____

_____

_____

ISBN: 9780170195997

3.5 Spherical lenses

Convex lenses (converging)
Rays parallel and close to the principal axis will, upon refraction, converge on the principal focus (F) of the lens.

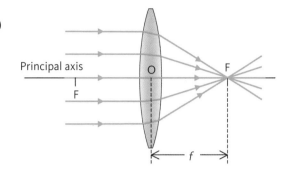

O Optical centre of the lens
F Principal focus (**real**)
f Focal length (**positive**)

Concave lenses (diverging)
Rays parallel and close to the principal axis will, upon refraction, form a diverging beam which appears to come from the principal focus (F) of the lens.

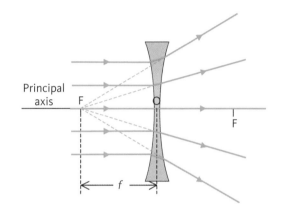

O Optical centre of the lens
F Principal focus (**virtual**)
f Focal length (**negative**)

Ray diagrams

The position, size and nature of the **image** (I) of an **object** (O) produced by a spherical lens can be found by drawing a ray diagram using three rules:

Rule 1 A ray parallel to the principal axis is refracted through the principal focus (or appears to diverge from it).

Rule 2 A ray through (or from) the principal focus is refracted parallel to the principal axis.

Rule 3 A ray incident on the optical centre of the lens passes through undeviated (without changing direction).

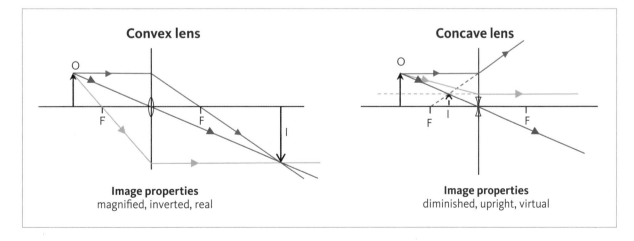

Any two rays are sufficient to determine the position of the image. A third ray can be drawn to confirm the answer.

ISBN: 9780170195997

Lens formula

The object and its image are related mathematically in terms of their position, size and the focal length of the lens.

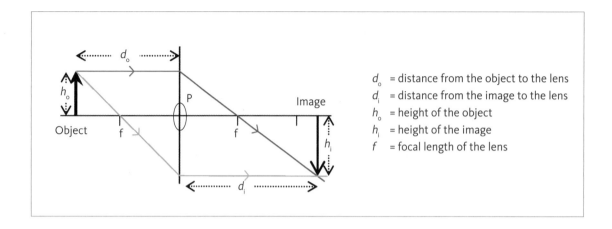

d_o = distance from the object to the lens
d_i = distance from the image to the lens
h_o = height of the object
h_i = height of the image
f = focal length of the lens

Magnification formula

To determine the magnification (m) produced by a lens:

$$m = \frac{d_i}{d_o} = \frac{h_i}{h_o}$$

Focal length formula

To determine the position of an image, an object or the focal length of a lens:

$$\frac{1}{f} = \frac{1}{d_i} + \frac{1}{d_o}$$

Real-is-positive convention

When solving lens formula problems it is essential to substitute the appropriate sign(s) into the equations. The value and sign of the answer will then provide information about its position, size and nature.

* Convex lenses have a **real** principal focus so the focal length (f) is **positive**.
* **Real** objects and images have **positive** distances.
* Concave lenses have a **virtual** principal focus so the focal length (f) is **negative**.
* **Virtual** objects and images have **negative** distances.

Magnifications (m) are always quoted as **positive** in questions, but the magnification of a virtual image should be substituted into equations as a **negative** value.

ISBN: 9780170195997

Worked example: Lens calculation

Determine the position and properties of the image formed by a 20 cm tall candle placed 66.0 cm away from a convex lens of focal length 11.0 cm.

Solution
Position and nature of the image:

Given	d_o	=	+66.0 cm
	f	=	+11.0 cm
Unknown	d_i	=	?
Equations	$\dfrac{1}{f}$	=	$\dfrac{1}{d_i} + \dfrac{1}{d_o}$
Substitute	$\dfrac{1}{11.0}$	=	$\dfrac{1}{d_i} + \dfrac{1}{66.0}$
Solve	$\dfrac{1}{d_i}$	=	$\dfrac{1}{11.0} - \dfrac{1}{66.0} = \dfrac{5}{66.0}$
	d_i	=	$\dfrac{66.0}{5}$
	d_i	=	**13.2 cm (3 s.f.)**

d_i is positive so image is **real**

Magnification:

Given	d_o	=	+66.0 cm
	d_i	=	+13.2 cm
	h_o	=	20 cm
Unknown	m	=	?
Equations	m	=	$\dfrac{d_i}{d_o}$
Substitute	m	=	$\dfrac{13.2}{66.0}$
Solve	m	=	0.200

$m < 1$ so image is **diminished**

And as	m	=	$\dfrac{h_i}{h_o}$ so $h_i = mh_o$
Hence	h_i	=	0.200×20
	h_i	=	**4.0 cm (2 s.f.)**

ISBN: 9780170195997

Exercise 3C

1 Complete the following ray diagrams by applying all three rules. (Scale: One square = 0.5 cm)

a Find the position and properties of the image formed by a 1.5 cm tall object placed 6.5 cm from a **convex** lens of focal length 3.5 cm.

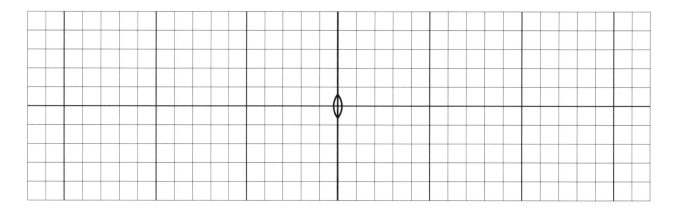

Properties of the image:

Nature: Size: Orientation: Magnification:

b Find the position and properties of the image formed by a 2.0 cm tall object placed 2.5 cm from a **concave** lens of focal length 7.5 cm.

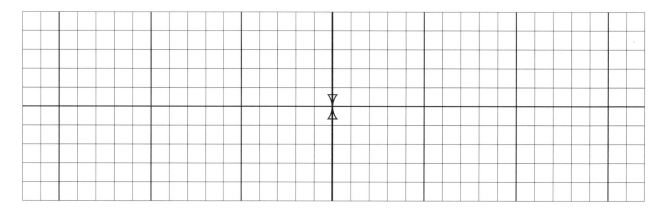

Properties of the image:

Nature: Size: Orientation: Magnification:

ISBN: 9780170195997

c Find the position and properties of the image formed by a 2.5 cm tall object placed 7.0 cm from a **convex** lens of focal length 3.5 cm.

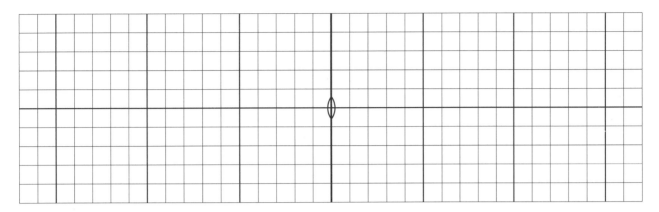

Properties of the image:

Nature: Size: Orientation: Magnification:

d The object and image are shown on the diagram below. State the nature of the image and then by drawing a ray diagram, determine the focal length of the lens.

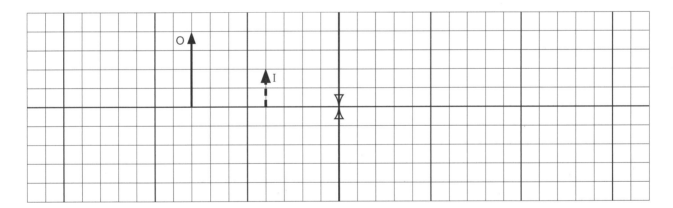

Properties of the image:

Nature: Focal length:

ISBN: 9780170195997

e The object and image are shown on the diagram below. State the nature of the image and then by drawing a ray diagram determine the focal length of the lens.

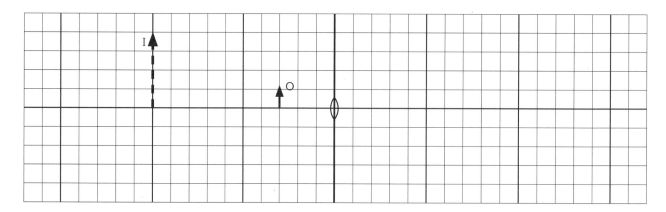

Properties of the image:

Nature: Focal length:

f Find the size and position of an object which forms a 0.5 cm tall virtual image in a **concave** lens. The image distance is 1.5 cm and the focal length of 2.0 cm.

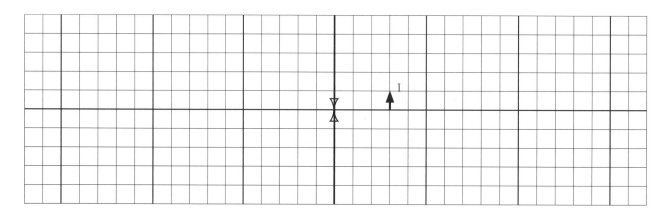

Properties of the image:

Size: Object distance:

ISBN: 9780170195997

2   Use the lens formula $\dfrac{1}{f} = \dfrac{1}{d_i} + \dfrac{1}{d_o}$ to solve the following problems.

a   Andrija places a convex lens 100 cm away from a lamp and focuses a real image onto a screen 36 cm away from the lens. Determine the focal length of the lens.

_____

_____

_____

_____

b   Using a convex lens of focal length 11 cm Devyani focuses an image of a candle onto a screen 66 cm away from the pole of the lens. Calculate the distance of the candle from the lens.

_____

_____

_____

_____

c   Vincent holds a concave lens 0.44 m away from a TV screen. The lens has a focal length of 5.0 m. Calculate the distance of the TV's virtual image from the pole of the lens.

_____

_____

_____

_____

3   Use $\dfrac{1}{f} = \dfrac{1}{d_i} + \dfrac{1}{d_o}$ and $m = \dfrac{d_i}{d_o} = \dfrac{h_i}{h_o}$ to solve the following problems.

a   Prachi uses a convex lens of focal length 73 cm to produce a real image of a super bright LED placed 74 cm from the pole of the lens. Given that the object is 0.8 cm tall, determine:
   i   the distance of the image from the lens
   ii   the magnification
   iii   the size of the image.

_____

_____

_____

_____

_____

_____

_____

_____

ISBN: 9780170195997

b Seung Jae can see a 150 m tall building from his house 2.30 km away. He uses a convex lens to form a real image of the building on a screen. The image is 0.75 m tall. Calculate:
   i the magnification of the image
   ii the distance of the image from the pole of the lens
   iii the focal length of the lens.

_____

_____

_____

_____

_____

_____

4 Use $\dfrac{1}{f} = \dfrac{1}{d_i} + \dfrac{1}{d_o}$ and $m = \dfrac{d_i}{d_o} = \dfrac{h_i}{h_o}$ to solve the following problems.

a Chirag is 1.72 m tall and stands 4.0 m away from a concave lens with a focal length of 12 m. Calculate:
   i the distance of Chirag's image from the lens
   ii the magnification
   iii the height of the image.

_____

_____

_____

_____

_____

_____

b Ginal is viewing a weta in a convex lens of focal length 77 cm. The weta's image is 12.6 cm long and appears to be 30.8 cm inside the lens. Calculate:
   i the distance of the weta from the pole of the lens
   ii the magnification of the image
   iii the weta's real length.

_____

_____

_____

_____

_____

ISBN: 9780170195997

5 Ethan uses a convex lens with a focal length of 35 cm to produce an image of a candle that is five times the size of the actual candle. Determine the two possible object distances where Ethan could place the candle to achieve these images.

_____

_____

_____

_____

_____

_____

_____

_____

6 Alex uses a convex lens of focal length 18 cm to produce an image on a screen of a lantern placed at a distance, d_o in front of the lens. Given that the image is one-third the size of the lantern, calculate the distance of the lantern from its image.

_____

_____

_____

_____

_____

_____

_____

7 Judy used a white light source and a thick magnifying glass to study the properties of images produced by convex lenses. When she looked closely at the images on a screen she noticed that they appeared to be coloured around the edges, as shown in the diagram below.

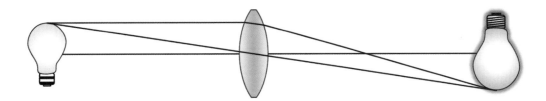

ISBN: 9780170195997

Judy thinks that the thick lens might be behaving like two prisms stuck together, as shown below.

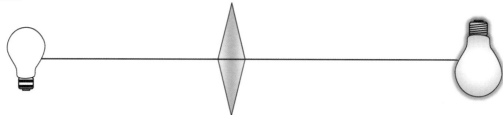

With the aid of the diagram above, explain why the image appears coloured.

_____

_____

_____

_____

_____

_____

_____

_____

_____

_____

8   Sue carries out an experiment to find the focal length of a convex lens. She places a candle at different distances from the lens and finds the location of the real image by moving a screen backwards and forwards until she achieves a sharp image. Her results are shown in the table.

Distance from the lens to the object, d_o (m)	Distance from the lens to the image, d_i (m)
0.20	No image
0.40	0.67
0.60	0.43
0.80	0.37
1.00	0.33
1.20	0.32

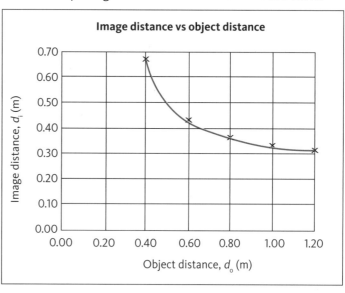

Sue doesn't recognise the shape of this graph, but knows that the relationship between image distance and object distance can be written as:

$$\frac{1}{f} = \frac{1}{d_i} + \frac{1}{d_o}$$

ISBN: 9780170195997

a Process the data by calculating $\dfrac{1}{d_o}$ and $\dfrac{1}{d_i}$ and putting the values in the table below.

(Note: You cannot process the data for 0.20 m as no image was found.)

b Write the unit for the processed data in the column heading.

$\dfrac{1}{d_o}$ ( )	$\dfrac{1}{d_i}$ ( )

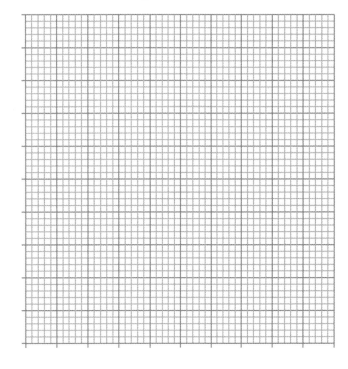

c Plot a graph with $\dfrac{1}{d_i}$ on the y-axis and $\dfrac{1}{d_o}$ on the x-axis.

d Determine the gradient and intercept of the graph and state the equation of the line.

_____

_____

_____

_____

e By comparing your equation of the line with the given formula, determine the focal length of the lens.

_____

_____

_____

_____

f Explain why Sue could not form an image on the screen when the object was placed 0.20 m from the lens.

_____

_____

_____

ISBN: 9780170195997

3.6 Wave phenomena

A **wave motion** will occur as a result of a regular disturbance of a mechanical medium, such as air (e.g. the transmission of sound); or an electric and magnetic field (e.g. transmission of light). It is a means of transferring energy from one region to another without there being any transfer of matter between the regions.

There are two general types of waves:

*   **Transverse waves:** Particles in the medium are displaced at right angles to the direction of travel of the wave (**s**ide-to-**s**ide vibrations).

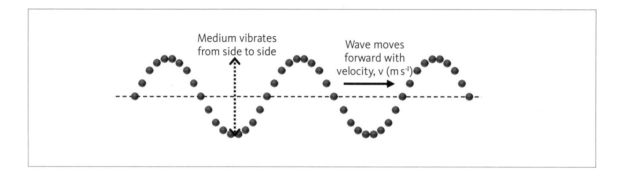

*   **Longitudinal** waves: Particles in the medium are displaced parallel to the direction of travel of the wave (**a long the same line**).

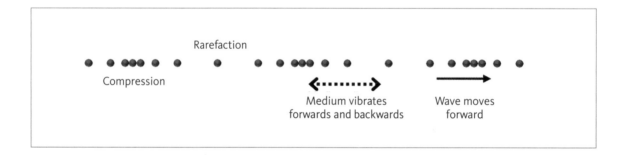

Representing waves

Both transverse and longitudinal waves can be represented as a sine wave with certain properties.

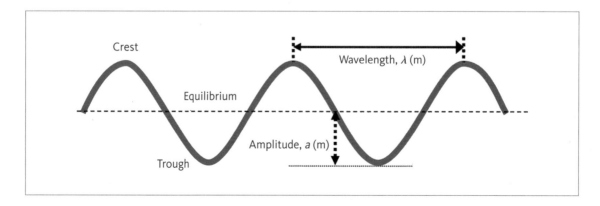

ISBN: 9780170195997

Wave properties

- **Crest** or peak: The top of a wave.

- **Trough**: The bottom of a wave.

- **Phase**: A means of comparing two points on a wave.
  - Two points are in phase if their maximum and minimum values occur at the same instant, otherwise there is said to be a phase difference.

- **Wavelength** (λ): The distance between two points on a wave with the same phase, measured in metres (m). Often measured in nm = 1 x 10^{-9} m for electromagnetic waves.

- **Amplitude** (a): The greatest displacement of the wave from its equilibrium position, measured in metres (m). The amplitude describes how much energy a wave transfers.

- **Period** (T): The time taken for one complete wave to pass a point, OR the time taken for a part of the wave to complete one oscillation, measured in seconds (s).

- **Frequency** (f): The number of complete waves produced each second, OR the number of complete waves to pass a point each second, measured in hertz (Hz) or per seconds (s^{-1}). Often measured in kHz = 1 x 10^3 Hz, MHz = 1 x 10^6 Hz, GHz = 1 x 10^9 Hz.

$$f = \frac{1}{T}$$

- Sound can also be described in terms of the pitch of the note, e.g. middle C, A sharp. High frequency waves have a high pitch. Low frequency waves have a low pitch.

Wave equation

Velocity, frequency and wavelength are related by the formula:

$$v = f\lambda$$

where: v = wave speed in m s^{-1}

f = frequency in hertz (Hz)

λ = wavelength in metres (m)

ISBN: 9780170195997

Exercise 3D

1   Ethan uses a spring to demonstrate wave motion. First, he sends a transverse wave down the spring, then he sends a longitudinal wave down the spring, as shown in the diagrams below.

    a   Complete the diagrams by filling in the labels.

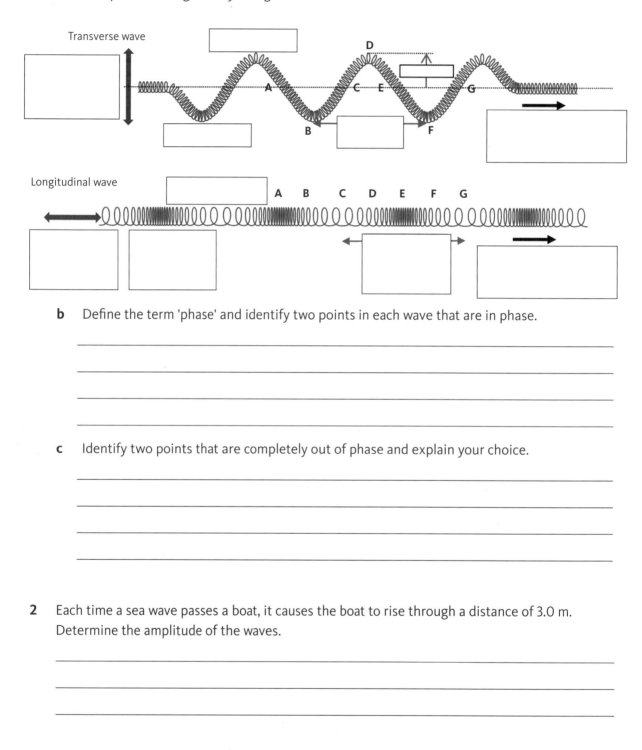

    b   Define the term 'phase' and identify two points in each wave that are in phase.

_____

_____

_____

_____

    c   Identify two points that are completely out of phase and explain your choice.

_____

_____

_____

_____

2   Each time a sea wave passes a boat, it causes the boat to rise through a distance of 3.0 m. Determine the amplitude of the waves.

_____

_____

_____

3   As Krishna stands on the deck of a large ship she observes that six waves fit perfectly between the front and the back of the ship.

    a   Given that the ship is 72 m long, determine the wavelength of the waves.

_____

ISBN: 9780170195997

b If it takes 30 seconds for the crest of a wave to travel from one end of the ship to the other calculate the speed of the wave.

_____

_____

c Calculate the frequency of the waves and state two possible units for frequency.

_____

_____

    i Frequency _____

   ii Unit _____

  iii Alternative unit _____

4 A speaker cone vibrates with a frequency of 440.0 Hz causing a sound wave to travel away from the speaker at 330.0 m s^{-1}.

a Define frequency, and explain how the sound wave travels through the air.

_____

_____

_____

_____

_____

_____

b Calculate the time period of the wave.

_____

_____

c Calculate the wavelength of the wave.

_____

_____

5 The label on a laser states that light of a wavelength of 550 nm is emitted. Given that light travels at a speed of 3.0×10^8 m s^{-1}, calculate:

a The frequency of the emitted light.

_____

_____

_____

b The time period of the waves.

_____

_____

_____

3.7   Waves and wavefronts

Wave diagrams can also be drawn by considering the waves as viewed from above. The crest of each wave is represented by a solid line, called a **wavefront**. Each wavefront is one wavelength apart. (The troughs are sometimes shown as dotted lines in between the crests.)

Circular wavefronts are produced by point sources, e.g. a stone dropped into water.

Plane wavefronts are produced by straight sources, e.g. a stick dropped into water.

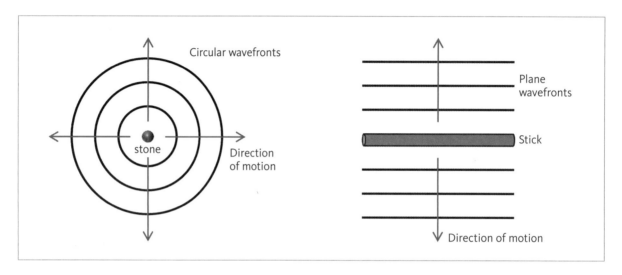

Diffraction

When waves pass through a gap they spread out into the region either side of the gap, and as the width of the gap decreases the amount of spreading increases.

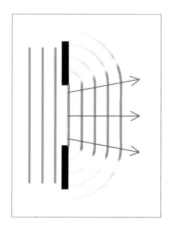
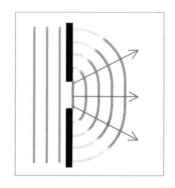
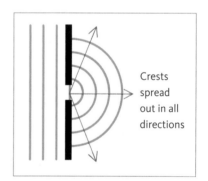

When the size of the gap is of a similar size to the wavelength the diffraction effects become very pronounced, producing semicircular wavefronts.

During diffraction the wavelength remains constant, but the amplitude decreases as the energy is spread out over a larger region.

(Note: When drawing diffraction diagrams, carefully measure the wavelength before the gap, and then mark about 4 additional wavelengths through the gap and beyond).

ISBN: 9780170195997

Exercise 3E

1 A laser emits a beam of light of frequency 5.00×10^{14} Hz. The beam spreads out slightly as it leaves the wide aperture (hole) at the end of the laser. The beam is then incident upon a very narrow slit.

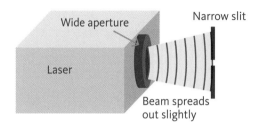

a Name the effect that causes the beam to spread out slightly as it leaves the end of the laser.

_____

b Complete the diagram above to show what happens to the waves after they pass through the very narrow slit.

c By considering the diagram, explain why the amount that the beam spreads out when it leaves the aperture of the laser is not the same as the amount the beam spreads out when it passes through the slit.

_____

_____

_____

2 A fireworks display is taking place at the harbour, but Jesse cannot see the display from her house. Jesse notices that she can hear the low pitch boom of the explosions but not the high pitched whistles and crackles. The two diagrams below show the low pitch waves, and the high pitch waves (next page) approaching the two tower blocks.

a Complete the diagrams to show what will happen when the waves pass through the gap.

Low pitch booming sounds

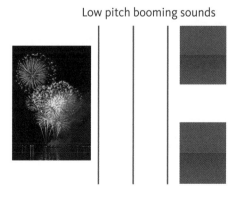

Jesse's house

High pitch whistling sounds

Jesse's
house

b Name the effect that allows Jesse to hear the low pitched sounds.

_____

c Explain why Jesse cannot hear the high pitched sounds at her house.

_____

_____

_____

_____

_____

3 Careena is staying in a hotel in a deep valley. She discovers that her mobile phone does not have any reception but the radio still works. The mobile phone transmits and receives using a microwave signal of wavelength of 13 mm, and the long wave radio signal has a wavelength of 130 m. With the aid of the diagram below, explain why the hotel can still receive long wave radio signals but not microwave phone signals.

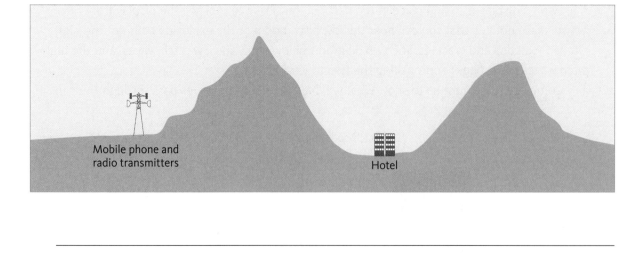

Mobile phone and
radio transmitters

Hotel

_____

_____

_____

_____

_____

ISBN: 9780170195997

Interference and superposition

When two or more pulses or waves are travelling in the same region they will **interfere** with each other, and the resultant wave will be the sum of the displacements of the contributing waves. The process of adding waves together is called **superposition**.

Constructive interference occurs if two pulses of similar wavelength and amplitude arrive in phase. The waves superpose to make a resultant wave with larger amplitude.

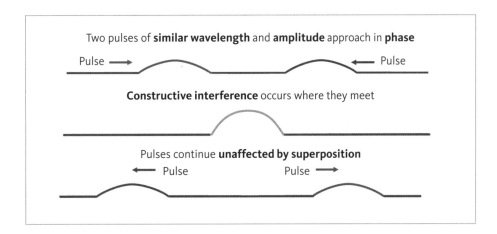

Destructive interference occurs if two pulses of similar wavelength and amplitude arrive out of phase. The waves superpose to make a resultant wave with smaller amplitude.

If the waves are identical but opposite in phase then the resultant wave will have no amplitude.

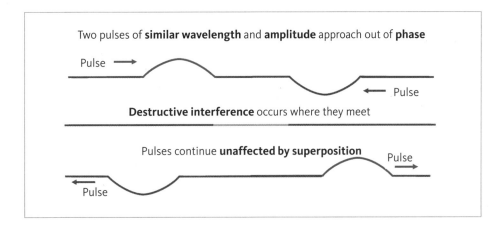

Standing waves

Standing waves are produced when two waves with the same amplitude, frequency and speed are travelling in opposite directions in the same region. When they meet, they interfere in such a way that some points along the wave do not move at all (called **nodes**), and other points move from a large positive amplitude to a large negative amplitude (called **antinodes**). The resultant wave does not appear to be travelling forward so it is called a **standing** or **stationary wave**.

ISBN: 9780170195997

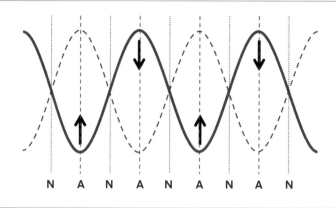

Antinodes (A) occur where:
- Large **A**mplitude.
- Waves meet **in phase**.
- **Constructive** interference.

Nodes (N) occur where:
- **N**o amplitude.
- Waves meet **out of phase**.
- **Destructive** interference.

Standing waves can be produced when a wave reflects from a surface and the transmitted wave interferes with the reflected wave. This occurs in springs, ropes and strings (e.g. a guitar string), in water or air (e.g. inside a musical instrument such as the hollow pipe of a flute).

Worked example: Superposition and constructive interference

Two pulses approach each other with a velocity of $1\,\text{m s}^{-1}$ along a spring of uniform strength, as shown in the diagram below (scale: one square = 1 m). Using superposition determine the resultant pulse produced by the two pulses after 3 s.

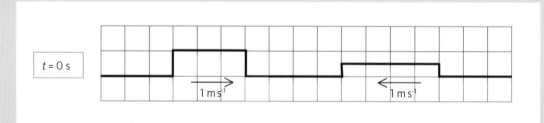

Solution

Step 1 *For each given time use the velocity to determine the position of the pulses, and faintly sketch in the pulses.*

Step 2 *Add the displacements together where the pulses overlap.*

Step 3 *Draw a single pulse (in blue), which is the sum of the two pulses. So if t = 3 s, if the pulses are travelling at $1\,\text{m s}^{-1}$ then the pulses will move 3 m, as shown in red and orange.*

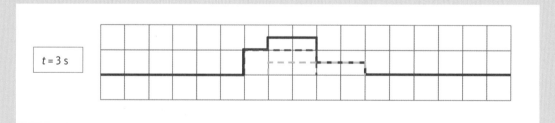

ISBN: 9780170195997

Exercise 3F

1   Two pulses approach each other with a velocity of 1 m s⁻¹ along a spring of uniform strength, as shown in the diagram below. (Scale: One square = 1 m)

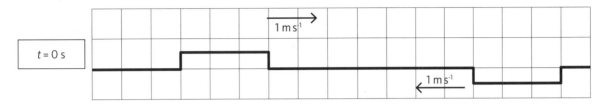

a   State the conditions necessary for destructive interference to occur.

_____

b   Using superposition, determine the resultant pulse produced by the two pulses after 4 s, 5 s and 8 s.

c   Explain what is observed at 5 s in terms of phase, and interference.

_____

_____

_____

_____

_____

2 Two pulses approach each other with a velocity of 1 m s⁻¹ along the surface of the sea as shown in the diagram below. (Scale: One square = 1 m)

a State the conditions necessary for constructive interference to occur.

_____

b Using superposition, determine the resultant pulse, produced by the two pulses after 4 s, 5 s and 7 s.

3 Sam is playing in the sink at home. He moves his hands up and down on the surface of the water, causing waves to travel across the surface of the water. Sam notices that when he moves both hands at either side of the sink, the waves no longer travel across the surface but just move up and down.

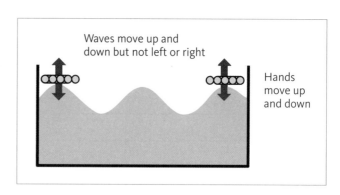

a Name the type of wave phenomena Sam has produced.

_____

b On the diagram above, identify:
   i An antinode, and label it **A**.
   ii A node, and label it **N**.

ISBN: 9780170195997

c State and explain the conditions necessary for this type of wave to occur.

_____

_____

_____

_____

_____

_____

2-point source interference

When plane waves pass through two narrow slits they are diffracted to produce semicircular wavefronts from each slit. These wavefronts spread out and interfere with each other constructively and destructively.

When light waves are used, the pattern can be observed on a screen, as a series of bright fringes (antinodes) and dark fringes (nodes).

To produce a clear pattern, the waves coming from the two slits must be coherent (similar wavelength, amplitude and in phase).

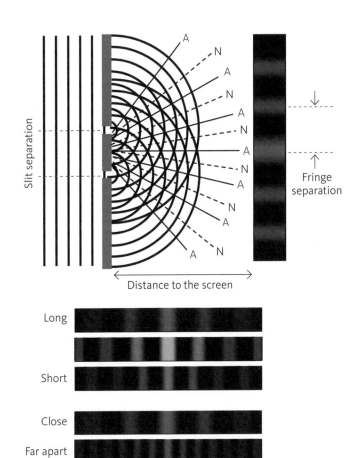

Factors that affect the diffraction pattern

- As the **wavelength** decreases, the distance between the fringes decreases.

- As the **slit separation** increases, the fringe separation decreases.

- As the **distance** to the screen decreases, the fringe separation decreases.

Exercise 3G

1 Heather is standing on top of a cliff looking down at the sea as it enters a natural harbour. The entrance to the harbour has a large rock in the middle. As the waves pass the entrance they spread out. The diagram below shows the wave crests approaching the entrance and passing into the harbour.

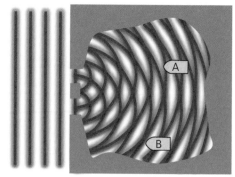

Natural harbour

a Name the phenomenon that causes the waves to spread out after they pass through the two entrances of the harbour, and state the conditions necessary for this effect to be obvious.

_____

_____

_____

_____

b Two boats are anchored in the harbour at positions A and B. Describe and explain the motion of each boat.

i Boat A

_____

_____

_____

_____

ii Boat B

_____

_____

_____

_____

2 A laser beam is shining on two narrow slits, causing a pattern of bright and dark fringes to appear on a screen, as shown in the diagram.

a Identify (name) the bright fringes and explain why they occur.

_____

_____

_____

_____

_____

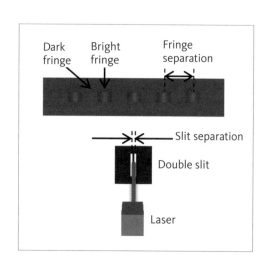

ISBN: 9780170195997

b State the necessary conditions for a clear pattern to be observed.

_____

_____

c Describe what effect increasing the **slits' separation** would have on the distance between the bright fringes.

_____

_____

d Describe what effect increasing the **frequency** would have on the distance between the bright fringes.

_____

_____

3 Two scientists want to explore the jungle to the south of their camp but are concerned about getting lost, so they set up two aerials at either end of their camp. Both aerials transmit the same frequency radio waves and they are in phase. When they have finished exploring, they switch on their radio receivers and start to walk in a straight line due west.

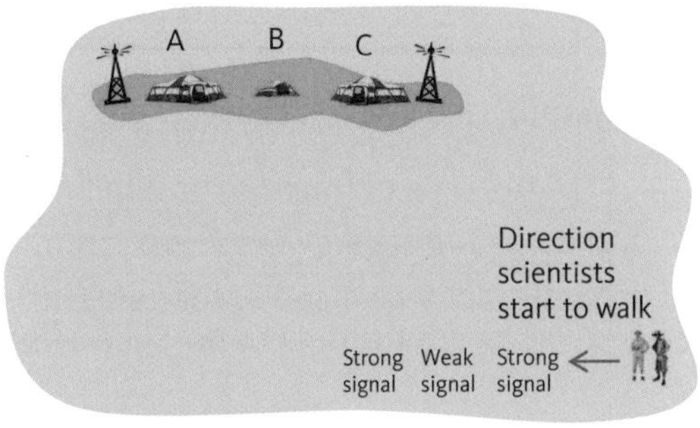

Direction scientists start to walk

Strong   Weak   Strong ←
signal   signal signal

a They notice that the radio signal gets stronger then weaker then stronger again. Explain why the signal changes strength as they walk along the path shown in the diagram.

_____

_____

_____

_____

_____

_____

_____

_____

Once they reach a strong signal they turn north and start walking. Each time the signal starts to become weaker they change direction so that they maintain a strong signal.

b Identify in front of which tent (A, B or C) the explorers leave the jungle, and explain how the distance between the strong and weak signals changes as they approach the camp.

_____

_____

_____

_____

_____

The scientists decide to test how changing the distance between the transmitters will affect the distance between the strong signals. They change the distance between the masts (d) and measure the distance between the central strong signal and the first strong signal (x) at a distance of 2.00 km from the camp. Their data is recorded in the table.

Transmitter separation, d (m)	Strong signal separation, x (m)	Processed data
10	60	
20	30	
30	20	
40	15	
50	12	

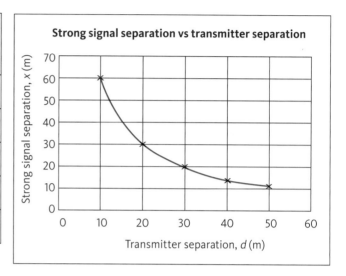

c Study the graph of strong signal separation (x) against transmitter separation (d), and state the type of the relationship between x and d.

_____

d Process the data to produce a linear relationship and plot the graph.

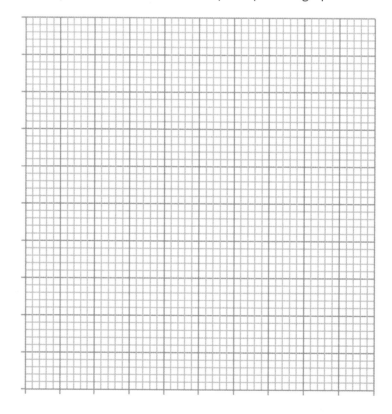

e Determine the gradient of the line and the intercept and state the equation of the line.

_____

ISBN: 9780170195997

f Given that the relationship between the transmitter separation (d) and the strong signal separation (x) for adjacent strong signals is given by the formula $x = \lambda L \cdot \dfrac{1}{d}$ use your equation of the line to calculate the wavelength of the signal being transmitted. (Distance from the towers to the explorers, L = 2.00 km)

_____

_____

_____

_____

_____

_____

3.8 Waves and boundaries

When a wave is incident upon a different medium it may cause the wave to reflect (bounce back) and/or refract (change speed and direction). The point where the two different mediums meet is referred to as a boundary, e.g. the surface of a glass window represents the boundary between air and glass; a harbour wall represents the boundary between water and stone.

Reflection at a plane boundary
Free end reflection of waves
When a wave is incident upon a boundary it may be reflected **without changing phase** if either:

* The boundary is a free end, or
* A slow moving wave is incident upon a medium in which it would travel faster.

This type of reflection can be seen when a pulse travels down a spring that is free to move side-to-side along a smooth rod. We refer to this as **free end reflection**.

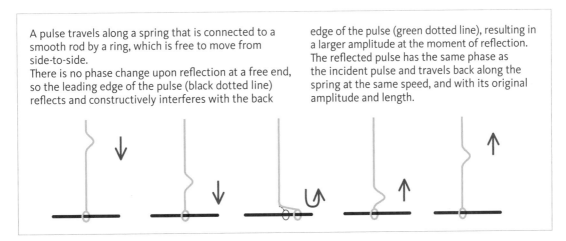

A pulse travels along a spring that is connected to a smooth rod by a ring, which is free to move from side-to-side.
There is no phase change upon reflection at a free end, so the leading edge of the pulse (black dotted line) reflects and constructively interferes with the back edge of the pulse (green dotted line), resulting in a larger amplitude at the moment of reflection. The reflected pulse has the same phase as the incident pulse and travels back along the spring at the same speed, and with its original amplitude and length.

When free end reflection is viewed from above, the crests of the waves approaching a surface reflect off and have the same phase, and the angle of the incident wavefront to the surface is the same as the angle of the reflected wavefront to the surface, as shown in the diagram at the top of the next page.

ISBN: 9780170195997

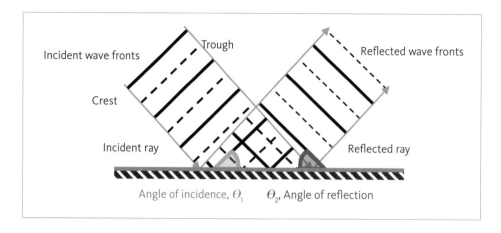

Fixed end reflection of waves

When a wave is incident upon a boundary it may experience **a 180° phase change** upon reflection, if either:

- The boundary is a fixed end, or
- If a fast moving wave reflects off a medium in which it would travel slower.

The same type of reflection can be seen when a pulse travels down a spring that is fixed in place. We refer to this as **fixed end reflection**.

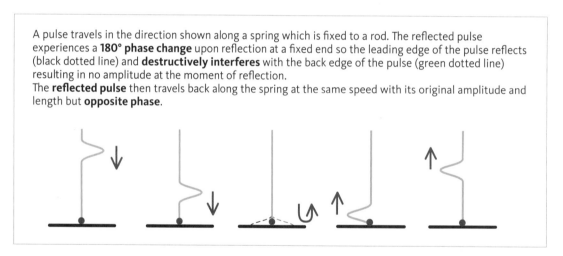

A pulse travels in the direction shown along a spring which is fixed to a rod. The reflected pulse experiences a **180° phase change** upon reflection at a fixed end so the leading edge of the pulse reflects (black dotted line) and **destructively interferes** with the back edge of the pulse (green dotted line) resulting in no amplitude at the moment of reflection.
The **reflected pulse** then travels back along the spring at the same speed with its original amplitude and length but **opposite phase**.

When fixed end reflection is viewed from above, the crests of the waves approaching a surface reflect off and have the opposite phase, however, the angle of the incident wave front to the surface is still the same as the angle of the reflected wave front to the surface, as shown in the diagram below.

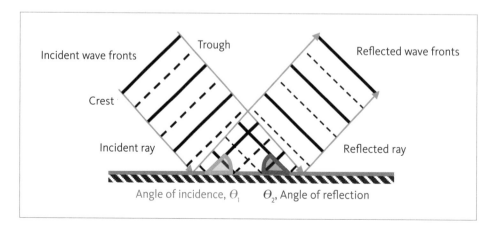

ISBN: 9780170195997

Exercise 3H

1 A pulse travelling at a velocity of 2 m s⁻¹ along a spring of uniform strength approaches a free end, as shown in the diagram below. (Scale: One square = 1 m)

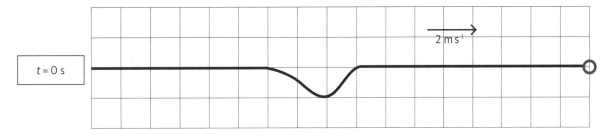

Draw a diagram to show the position, shape and phase of the pulse after 6 s.

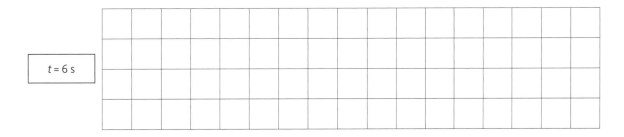

2 A wave travelling at a velocity of 2 m s⁻¹ along a spring of uniform strength approaches a fixed end, as shown in the diagram below. (Scale: One square = 1 m)

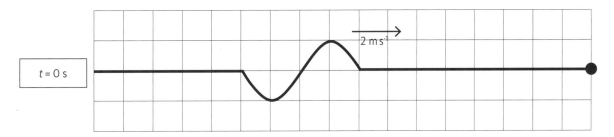

Draw a diagram to show the position, shape and phase of the pulse after 5 s and 7 s. Use superposition if required.

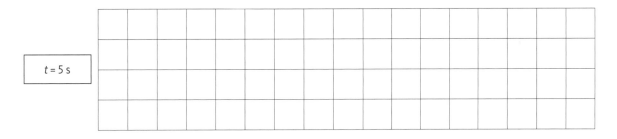

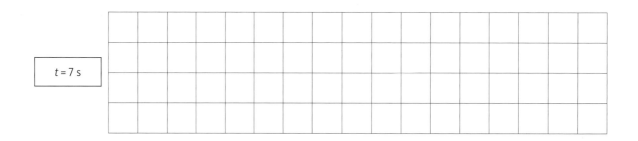

$t = 7$ s

3 Two pulses, both travelling at a velocity of 3 m s⁻¹ to the right along a spring of uniform strength, approach a free end, as shown in the diagram below. (Scale: One square = 1 m)

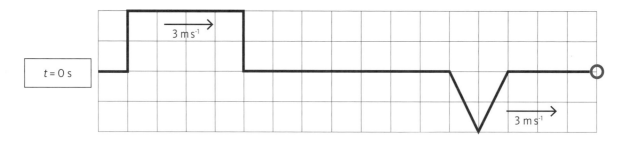

$t = 0$ s

3 m s⁻¹

3 m s⁻¹

Draw a diagram to show the position, shape and phase of the pulse after 2 s and 3 s. Use superposition if required.

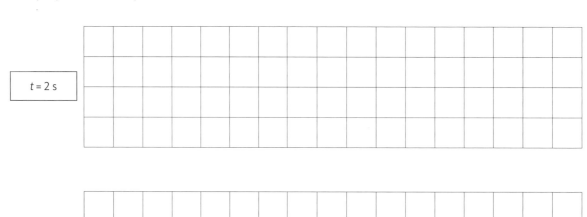

$t = 2$ s

$t = 3$ s

ISBN: 9780170195997

4 A sea wave is observed travelling towards the foot of a cliff at an angle, as seen from above in the diagram.

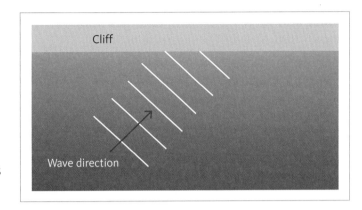

Cliff

Wave direction

a Complete the diagram to show the reflected wavefronts for the waves shown, and draw on three more complete wavefronts to show where the reflected waves go. Water waves do not experience a **phase change** when they reflect.

b Explain what is meant by the term phase change, and hence describe and explain what will happen to the amplitude of the resultant wave as it strikes the cliff face.

_____

_____

_____

_____

_____

Refraction at a plane boundary

When a wave enters a new medium in which it travels at a different speed it will experience a change in the wavelength of the wave, but the frequency will remain constant. If the wave is incident upon the boundary at an angle, then the wave will also change its direction of motion. This phenomenon is called refraction.

The end of each wavefront closest to the boundary enters the new medium and slows down. The other end of the wavefront is still moving quickly. This causes the wavefront to:

- **change direction**, and
- **change wavelength**

but the **frequency remains constant**:

$$\frac{\sin \Theta_1}{\sin \Theta_2} = \frac{v_1}{v_2} = \frac{\lambda_1}{\lambda_2} = \frac{n_2}{n_1}$$

where $\Theta_{1/2}$ = angle of incidence/angle of refraction
 $v_{1/2}$ = speed in medium 1/medium 2
 $\lambda_{1/2}$ = wavelength in medium 1/medium 2
 $n_{1/2}$ = refractive index in medium 1/medium 2

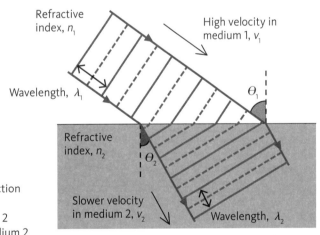

Refractive index, n_1

High velocity in medium 1, v_1

Wavelength, λ_1

Θ_1

Refractive index, n_2

Θ_2

Slower velocity in medium 2, v_2

Wavelength, λ_2

Phase and wave parameter changes upon refraction

When a wave is incident upon a medium in which it travels at a different speed we must also consider what happens to its **phase**.

Travelling from a 'fast' to a 'slow' medium

When the pulse reaches the boundary, it is reflected and transmitted.

The 'slow' medium acts like a 'fixed end' so the reflected pulse is phase changed by 180° and travels back through the fast medium at the same speed and with the same wavelength.

The transmitted pulse slows down but is not phase changed. This causes the wavelength of the transmitted pulse to decrease, and it will be closer to the boundary compared to the reflected pulse.

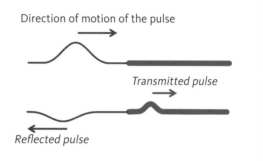

Travelling from a 'slow' to a 'fast' medium

When the pulse reaches the boundary, it is reflected and transmitted.

The 'fast' medium acts like a 'free end' so the reflected pulse is NOT phase changed and travels back through the slow medium at the same speed with the same wavelength.

The transmitted pulse speeds up but is not phase changed. This causes the wavelength of the transmitted pulse to increase, and it will be further from the boundary compared to the reflected pulse.

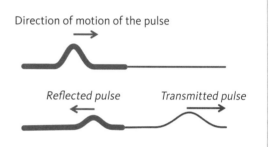

Worked example: Light wave frequency

As Isabelle is swimming in her pool one night she wonders if the decorative coloured lights will look different when she is underwater.

If each red light emits light of wavelength 660 nm calculate the wavelength of the light underwater, given that light travels at 3.00×10^8 m s^{-1} in air and 2.25×10^8 m s^{-1} in water.

Solution

$Given \qquad \lambda_a = 660 \times 10^{-9}$ m

$\qquad\qquad\quad v_a = 3.00 \times 10^8$ m s^{-1}

$\qquad\qquad\quad v_w = 2.25 \times 10^8$ m s^{-1}

$Unknown \quad \lambda_w = ?$

$Equations \quad \dfrac{v_a}{v_w} = \dfrac{\lambda_a}{\lambda_w}$

$Substitute \quad \dfrac{3.00 \times 10^8}{2.25 \times 10^8} = \dfrac{660 \times 10^{-9}}{\lambda_w}$

$Solve \qquad \lambda_w = \dfrac{660 \times 10^{-9}}{1.33}$

$\qquad\qquad\quad \lambda_w = 496 \times 10^{-9}$ m

ISBN: 9780170195997

Exercise 3I

1   The diagram below shows wavefronts of wavelength 12 m approaching a shallow region of water. The waves are travelling at 3.0 m s^{-1} in the deep region, but slow down to 1.5 m s^{-1} in the shallow region.

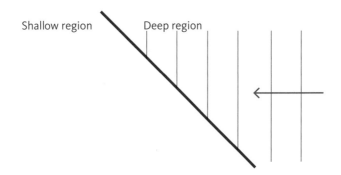

Shallow region   Deep region

a   On the diagram above, draw in the refracted wavefronts in the shallow region and an arrow to show the new wave direction.

b   Calculate the frequency of the waves in the deep water, and describe what happens to the frequency when the waves enter the shallow region.

_____

_____

_____

_____

c   Calculate the wavelength of the waves in the shallow water.

_____

_____

2   A swimming pool used for learning to SCUBA dive has a shallow section that is 1.0 m deep, and a deep section that is 4.0 m deep, separated by a boundary where there is a sudden change in height. A diver is paddling her feet in the water 30.0 times each minute, sending waves travelling towards the boundary. She observes that the waves change direction when they pass from one section to the other, as shown in the diagram below.

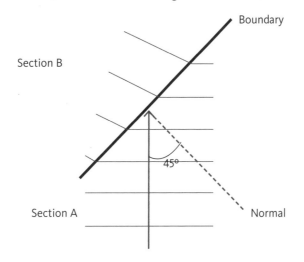

Boundary

Section B

45°

Section A   Normal

ISBN: 9780170195997

a Draw an arrow on the diagram at the bottom of the previous page to show the new direction of the waves in Section B.

b State which section (A or B) is deeper and explain your choice.

_____

_____

_____

c Show that the frequency of the waves is 0.5 Hz.

_____

_____

d Determine the angle of refraction if the incident wave strikes the boundary at an angle of 45° at a speed of 0.75 m s⁻¹, and leaves the boundary with a wavelength of 2.0 m.

_____

_____

_____

_____

_____

_____

3 A pulse travelling along through a slow medium approaches a faster medium. Draw a diagram to show what the pulse will look like shortly after it strikes the boundary, showing the position, shape and phase of the reflected and transmitted pulses.

Before

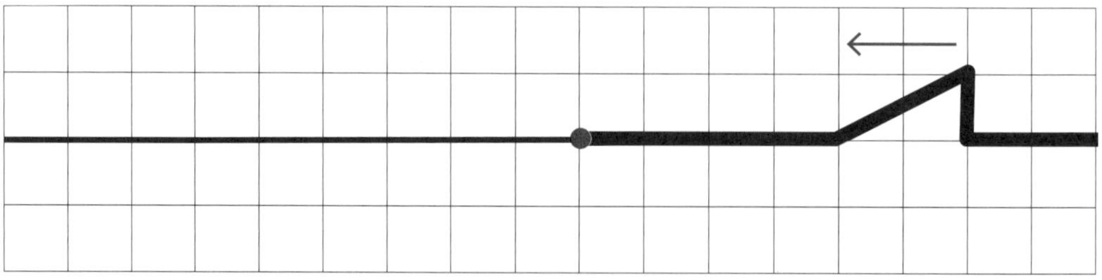

After

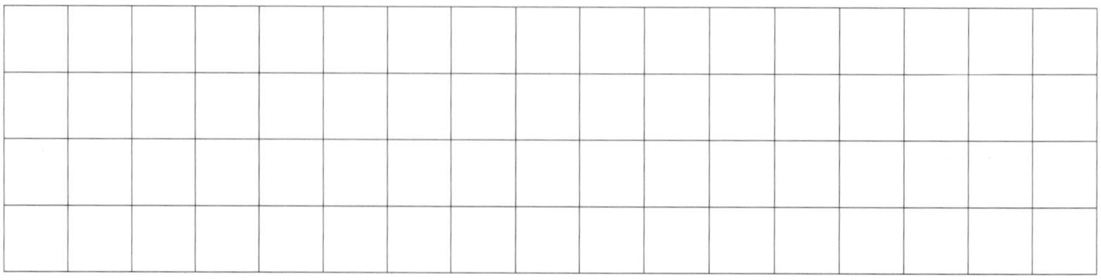

ISBN: 9780170195997

4 Two springs are connected together. One spring is light and waves travel at 3.0 m s^{-1} through it. The other spring is heavier causing waves to travel at 2.0 m s^{-1}. A pulse is sent along the light spring towards the heavy spring, as shown below. (Scale: One square = 1.0 m)

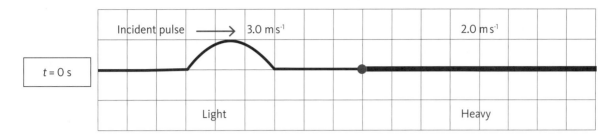

a Using information in the diagram, calculate the length of the transmitted pulse.

_____

_____

b Draw a diagram to show the relative position, shape and phase of the reflected and transmitted pulses 2 s after it hits the boundary.

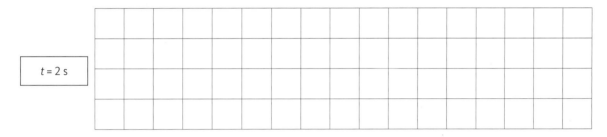

5 Two springs are connected together. One spring is heavy and waves travel slowly through it, while the other spring is lighter causing waves to travel faster. A wave is sent along the heavy spring towards the light spring, as shown below. Draw a diagram to show the position, shape and phase of the reflected and transmitted waves after it hits the boundary.

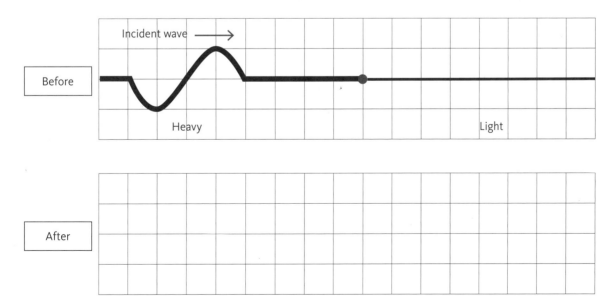

4 Mechanics

4.1 Vectors

Scalars and vectors

A **scalar quantity** has only magnitude (size or scale). Scalar quantities should always be written with a quantity symbol, a size and a unit, e.g. distance $d = 10$ m and speed $v = 14$ m s^{-1}.

A **vector quantity** has both magnitude and direction. Vector quantities should always be written with an <u>underline</u> below the quantity symbol, a size, a unit and a direction, e.g. displacement $\underline{d} = 42$ m (W) and velocity $\underline{v} = 20$ m s^{-1} ($\rightarrow$). Vectors can be represented either by:

- an underlined letter, e.g. $\underline{v}$
- two letters representing each end of the vector with an arrow above, e.g. $\overrightarrow{AB}$
- or by using **bold script**, e.g. **F**.

Drawing vectors

As vectors are geometric quantities they should always be drawn:

- to **scale** with a ruler
- with an **arrow** to show their direction
- with a **quantity symbol** to identify the type of vector.

An accurately drawn vector diagram can be used to solve vector problems instead of using vector mathematics.

Adding vectors

When adding vectors that are not in a straight line, we find the resultant by forming a **vector triangle**.

The size of the resultant vector (R) can be found by using Pythagoras' theorem. (Note: P, Q and R do not have arrows over them as we are only using their sizes.)

$$R^2 = P^2 + Q^2$$

The direction of the resultant vector (Θ) can be found by using trigonometry. The angle at the tail of the resultant vector should be chosen and clearly labelled on the diagram.

$$\tan \Theta = \left(\frac{P}{Q}\right) \quad \text{so} \quad \Theta = \tan^{-1}\left(\frac{P}{Q}\right)$$

ISBN: 9780170195997

Subtracting vectors *(change)*

In order to find the size and direction of the change in velocity ($\Delta \underline{v}$), we must first write a vector equation for change:

$$\Delta \underline{v} = \underline{v}_f - \underline{v}_i$$

For example:

$$\Delta \underline{v} = \underline{D}(\downarrow) - \underline{C}(\rightarrow)$$

and then reverse the sign and direction of the initial vector:

$$\Delta \underline{v} = \underline{D}(\downarrow) + \underline{C}(\leftarrow)$$

Now draw a **vector addition diagram** with the vectors positioned tail-to-head.

The size of the change vector (Θ) can be found using Pythagoras' theorem:

$$\Delta v^2 = D^2 + (-C)^2$$

The direction of the change vector (Θ) can then be found by trigonometry. As a general rule, the angle to the tail of the resultant vector should be found.

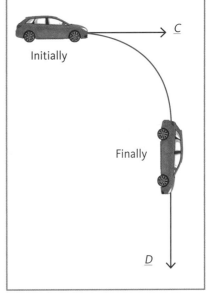

Initially

Finally

$\underline{D}$

Change

$\Delta \underline{v}$

Θ

$\underline{D}$

$-\underline{C}$

Resolving vectors *(components)*

A single vector can be split into two vectors perpendicular (90°) to each other. The process of finding the two perpendicular components of a single vector is called **resolving** a vector.

Consider a child's toy being pulled at an angle by a force ($\underline{F}$). The single vector can be resolved into horizontal ($\underline{F}_h$) and vertical ($\underline{F}_v$) components using trigonometry.

$$\sin \Theta = \frac{F_v}{F} \quad \text{hence} \quad F_v = F \sin \Theta \quad \textbf{and} \quad \cos \Theta = \frac{F_h}{F} \quad \text{hence} \quad F_h = F \cos \Theta$$

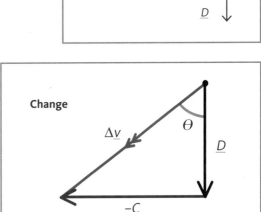

$\underline{F}$

$\underline{F}$

$\underline{F}_v$

Θ

$\underline{F}_h$

ISBN: 9780170195997

Exercise 4A

Vector addition

1   Draw a scale vector diagram to determine the size and
    direction of the resultant vector for the following
    situations.

   a   Two people pushing a toboggan chair with a force
          of 12 N east and 18 N east. Find the size and direction
          of the resultant force.

   _____

   _____

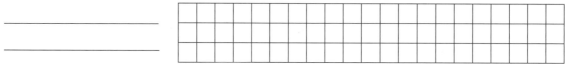

   b   A journey involving a displacement of 85 m west followed by a displacement of 55 m east.
          Find the size and direction of the resultant displacement.

   _____

   _____

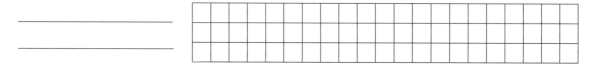

   c   Walking with a velocity of 3.0 m s⁻¹ south on a travelator moving with velocity of 2.0 m s⁻¹
          north. Find the size and direction of the resultant velocity.

2   Draw a vector diagram to determine the size and direction of the resultant vector for the
    following situations. (*Hint: The vectors are perpendicular to each other, so you need to use
    Pythagoras' theorem and trigonometry.*)

   a   Two tug boats pulling a large ocean liner with a force of 1600 N east and 1200 N south. Find
          the size and direction of the resultant force.

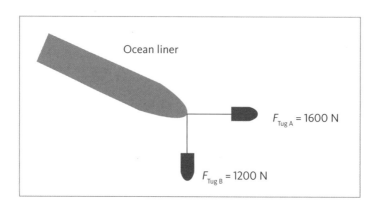

Ocean liner

$F_{Tug\,A}$ = 1600 N

$F_{Tug\,B}$ = 1200 N

ISBN: 9780170195997

_____

_____

_____

_____

_____

_____

b A car drives 120 km west then turns north and drives a further 50 km. Find the size and direction of the resultant displacement.

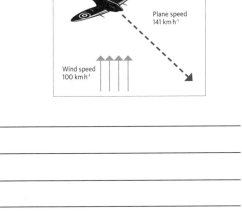

_____

_____

_____

_____

_____

_____

_____

c An aircraft flies south–east (bearing of 135°) with a velocity of 141 km h⁻¹ but there is a crosswind blowing north at 100 km h⁻¹ throughout the flight. Find the size and direction of the resultant velocity.

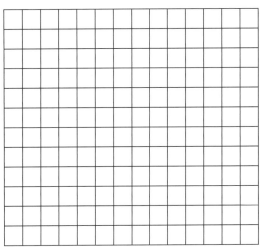

_____

_____

_____

_____

_____

Vector subtraction

3   Draw a vector diagram to determine the size and direction of the change vector for the following situations.

   a A water skier is floating in the water at the beginning of a race. The speed boat initially pulls the skier with a force of 440 N east, but once out of the water the force decreases to 120 N east. Find the size and direction of the change in force.

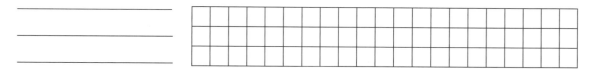

   b A car drives 200 km west on the first day of a journey. By the end of the second day the car is 300 km west from where it started. Find the size and direction of the change in displacement from day 1 to day 2.

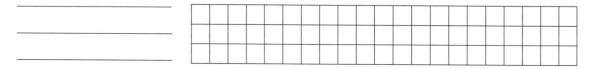

   c A cyclist initially travelling east with a velocity of 6.0 m s^{-1} turns around and cycles west with velocity of 10.0 m s^{-1}. Find the size and direction of the change in velocity.

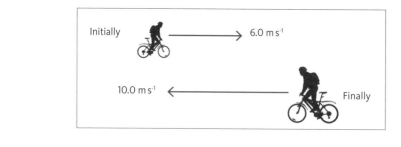

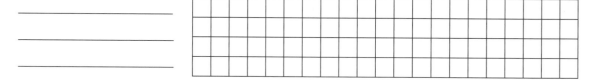

4   Draw a vector diagram to determine the size and direction of the change vector for the following situations. *(Hint: The vectors are perpendicular to each other, so you need to use Pythagoras' theorem and trigonometry.)*

   a A football, initially travelling at 3.0 m s^{-1} south, is kicked by a striker so that it travels at 6.0 m s^{-1} west. Find the size and direction of the **change** in velocity.

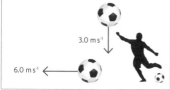

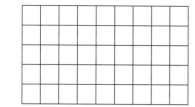

ISBN: 9780170195997

b A car driving east at 40 km h⁻¹ approaches a bend in the road. The car leaves the bend travelling at 32 km h⁻¹ north. Find the size and direction of the change in velocity.

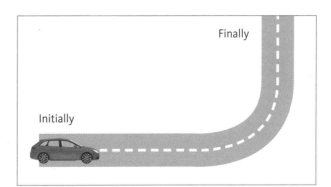

_____

_____

_____

_____

_____

_____

_____

Resolving vectors

5 Resolve the following vectors into horizontal and vertical components and complete the vector diagram.

a A shot put is thrown at an angle of 40° to the horizontal with a force of 200 N. Resolve the vector to find the horizontal and vertical force components.

_____

_____

_____

_____

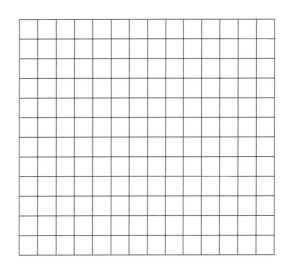

b A javelin hits the ground at a speed of 18 m s⁻¹ at an angle of 30° to the horizontal. Resolve the vector to find the horizontal and vertical velocity components at the moment of impact.

_____

_____

_____

_____

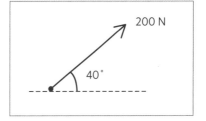

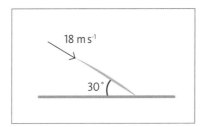

6 Resolve the following vectors into components parallel and perpendicular to the plane and complete the vector diagram.

a A cyclist is climbing a hill at an angle of 10° to the horizontal at a steady speed. The combined weight of the cyclist and the bike is 880 N.

$F_g = 880\ N$

i Resolve the weight vector to find the force components parallel and perpendicular to the road, and draw them on the diagram.

_____

_____

_____

_____

_____

_____

ii How much output force must the cyclist produce to keep the bike going up the hill at a steady speed? (*Hint: The forces must be balanced for an object to move at a steady speed.*)

_____

_____

iii Draw the cyclist's output force vector on the diagram.

$F_g = 640\ N$

b A skier and her equipment (total weight 640 N) is travelling down a slope at an angle of 40° to the horizontal at a steady speed.

i Resolve the weight vector to find the force components parallel and perpendicular to the slope.

_____

_____

_____

_____

_____

_____

ii How much support force is the snow providing to stop the skier from sinking?

_____

_____

iii Draw the support force vector on the diagram.

ISBN: 9780170195997

4.2 Forces and their effects

Forces

A force can cause an object to change its speed, direction, spin or shape. All forces are measured in **newtons** (N). Whenever an object experiences a **push** or a **pull** it is due to the action of a force.

Free body force diagram

Forces are represented by vectors, with a size, a direction and a label.

A free body force diagram shows all the forces acting on a single object. Consider an apple hanging from a tree: The weight force due to Earth's gravity pulls the apple towards the ground, and the support force from the branch prevents it from falling. In the diagram we do not consider the force of the apple on the branch, or the force of the apple's gravity pulling the Earth upwards. (Yes, it really does! See Newton's Third Law.)

Support, F_s

Weight, F_g

Contact and non-contact forces

Forces can be grouped into two categories: non-contact and contact forces.

Non-contact forces act on objects even though they are not in contact with each other, for example, the force due to gravity on objects with mass, the magnetic force on ferromagnetic materials and the electrostatic force between charged objects. The region in which that force is felt is described as a **field**, for example, a gravitational field is a region in space in which objects with mass experience a force of attraction.

If the force occurs as a result of something touching an object it is described as a contact force, for example, the force of friction between two surfaces in contact, tension due to an elastic band, the thrust of a rocket or jet engine or the upthrust from the sea causing a boat to float.

Combining force vectors

Force vectors can be added, subtracted or resolved using vector mathematics (see pages 88–89).

4.3 Newton's laws

Newton's First Law – balanced forces

If the net force acting on an object is **zero**, then either:
- a stationary object will remain stationary
- or a moving object will continue moving at a steady speed in a straight line (i.e. at a **constant velocity**).

Newton's Second Law – unbalanced forces

If the net force acting on an object is **not zero**, then either:
- a stationary object will start moving
- or a moving object will change its speed and/or change direction (i.e. **change velocity**).

ISBN: 9780170195997

When an object changes its velocity (speed or direction of motion), it can be said to have **accelerated**. Therefore:

net force (N)	=	mass (kg)	x	acceleration (m s^{-2})

Expressed mathematically:

$$\underline{F}_{net} = m\underline{a}$$

My FAVoUriTe questions

Any question involving Newton's First and Second Laws can be considered in terms of six key ideas: **m**ass, **F**orce, **A**cceleration, **V**elocity, **u**nbalanced/balanced and **t**ime. The order of the three central quantities is important, as each leads into the next. So discussions should be worded accordingly:

Force → **A**cceleration → **V**elocity

or:

Velocity → **A**cceleration → **F**orce

The ideas about mass, time and whether the forces are balanced or unbalanced provide additional clarity to the discussion.

Newton's Third Law

When object A exerts a force on object B, then object B will exert an equal but opposite force on object A.

Consider a box sitting on a table. The box exerts a force downwards on the table. The table exerts an equal sized force on the box but in the opposite direction. This is referred to as an **action-reaction pair**.

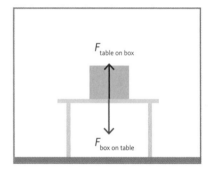

Mass and weight

Mass can be defined as a measure of an object's resistance to changes in its motion, or a measure of the **intertia** of an object. The unit of mass (m) is the kilogram (kg).

On Earth, if we ignore air resistance, every falling object experiences a uniform acceleration of $g = 9.81$ m s^{-2} downwards, due to the gravitational force of attraction between the mass of the Earth and the mass of the object. This force due to gravity (F_g) is more commonly described as the **weight** of an object.

Using Newton's Second Law, $\underline{F}_{net} = m\underline{a}$, so the weight can therefore be determined by:

weight (N)	=	mass (kg)	×	acceleration due to gravity (m s^{-2})

Expressed mathematically:

$$\underline{F}_g = mg$$

ISBN: 9780170195997

Worked example: Unbalanced forces

Ian is at cricket practice where they are teaching the players to 'follow through' after each shot. This technique involves hitting the ball with the same force but keeping the bat and ball in contact for longer. Explain what effect this will have on the ball's motion.

Solution

Given	'same force' implying Newton's laws, and 'time'
Unknown	'motion' which implies velocity and acceleration
Equations	$\underline{F}_{net} = m\underline{a}$ and $\underline{a} = \frac{\Delta \underline{v}}{\Delta t}$ with acceleration as the linking idea
Substitute	**M**y **FAV**o**U**ri**Te**: $\underline{F}_{net} = m\underline{a}$ so $\Delta \underline{v} = \underline{a}\Delta t$
Solve	The **M**ass of the ball and the **F**orce acting on it stay the same, so the **A**cceleration of the ball must also remain constant as $\underline{F}_{net} = m\underline{a}$. The same **A**cceleration is acting on the ball but for a longer **T**ime, and this will result in a greater change in **V**elocity as $\Delta \underline{v} = \underline{a}\Delta t$. This means that the ball will leave the bat moving with a greater velocity.

Note: This problem could have been solved in terms of momentum instead.

Exercise 4B

1   Describe what is meant by the term 'force', and explain the difference between a contact and a non-contact force.

_____

_____

_____

_____

_____

_____

2   Explain what effect a non-zero net force will have on the motion of an object using **both** Newton's First and Second law.

_____

_____

_____

_____

_____

_____

_____

ISBN: 9780170195997

3   Complete the following free body force diagrams to identify the forces acting in each of the
    following situations, and describe the forces as balanced or unbalanced. Explain your reasons.
    Each force vector should be drawn with a relative size and direction based on the information
    provided.

   a   A helicopter hovering in mid-air.

   _____

   _____

   _____

   _____

   _____

   _____

   b   An aircraft increasing its speed while flying straight and level.

   _____

   _____

   _____

   _____

   _____

   _____

   c   A bungee jumper at the bottom of a bounce.

   _____

   _____

   _____

   _____

   _____

   d   A go-cart travelling at a steady speed down a hill.

   _____

   _____

   _____

   _____

   _____

   _____

ISBN: 9780170195997

4 Harry (of mass 80 kg) and Sally (of mass 50 kg) are at the ice rink facing each other when Sally suddenly pushes Harry. Explain what will happen using both Newton's Second and Third laws.

_____

_____

_____

_____

_____

_____

_____

5 Use **_F = ma_** to solve each of the following situations.

a A cyclist and her bike have a combined mass of 75 kg. If she accelerates at 1.8 m s^{-2} calculate the net force acting on her and the bike.

_____

_____

_____

b A net force of 700 N acts on a skydiver causing him to accelerate at 9.8 m s^{-2}. Calculate the mass of the skydiver.

_____

_____

_____

c A mouse of mass 50 g jumps with a net force of 0.30 N. Calculate the size of the acceleration.

_____

_____

_____

_____

_____

6 Use $\underline{F} = m\underline{a}$ and $\underline{a} = \dfrac{\Delta v}{\Delta t}$ to solve each of the following situations.

    a When a raindrop hits the ground a force of 4.5×10^{-3} N is applied to it, causing it to decelerate from a speed of 9.09 m s^{-1} to 0.00 m s^{-1} in 1.01 seconds. Determine the acceleration of the raindrop and hence the mass of the rain drop.

_____

_____

_____

_____

_____

    b A sprinter of mass 88 kg produces an average net force horizontally of 176 N. Determine his acceleration during a race taking 10 s and thereby calculate his final speed.

_____

_____

_____

_____

_____

    c Explain why a tennis player can increase the velocity that the ball leaves the head of the racquet during a serve using the same force but by keeping the racquet in contact with the ball for a longer time.

_____

_____

_____

_____

_____

_____

_____

_____

ISBN: 9780170195997

d During a school fair, the students set up a stall to throw wet sponges at teachers who have been locked in the 'stocks'. A student throws two sponges, one with double the water in it compared to the other sponge. Explain why the sponge that contains double the water will have a greater force on impact, even though both sponges are thrown at the same speed.

_____

_____

_____

_____

_____

_____

_____

_____

_____

7 Use $F_{net} = \Sigma F$ and $F = ma$ to solve each of the following situations.

a A car of mass 950 kg accelerates at 0.80 m s^{-2}. Determine the size of the net force acting on the car, and calculate the size of the friction force acting on the car given that the engine provides a driving force of 1500 N.

_____

_____

_____

_____

_____

b A rocket of weight 2.95 × 10^3 N fires its engines, providing 3.55 × 10^3 N of thrust. Determine the size of the net force and hence calculate the acceleration of the rocket (g = 9.8 N kg^{-1}).

_____

_____

_____

_____

_____

ISBN: 9780170195997

8 Draw labelled action-reaction pair force vectors on the following diagrams, for each of the objects shown. The length of the force vectors should show how the forces result in the motion described in each situation.

a The moon orbiting the Earth.

b A helicopter accelerating upwards.

ISBN: 9780170195997

9 An 8.0 kg box is accelerated at 3.9 m s^{-2} along a table by a 12.0 kg mass hanging over the edge of the table, as shown in the diagram.

    a Draw free body diagrams of the forces on the box and the forces on the mass.

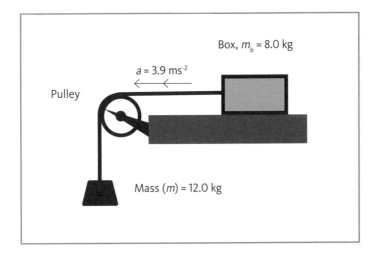

Box, m_b = 8.0 kg

a = 3.9 ms^{-2}

Pulley

Mass (m) = 12.0 kg

    b Determine the size of the net force acting on the box, and therefore calculate the size of the friction force between the table and the box.

_____

_____

_____

10 Tracey is a passenger in a car as it accelerates away from traffic lights. Explain why Tracey feels as though she is being pressed into her seat as the car accelerates.

_____

_____

_____

_____

_____

_____

11 Maddie is standing in an elevator waiting to travel up to the top floor. As the elevator sets off she notices that she feels heavier than normal.

    a Explain why Maddie feels heavier as the elevator sets off upwards.

_____

_____

_____

_____

_____

_____

_____

ISBN: 9780170195997

b Describe how Maddie will feel when the elevator is moving at a steady speed and explain your answer.

_____

_____

_____

_____

_____

_____

4.4 Momentum

When an object with mass (m) is moving with velocity (v) it possesses a certain 'quantity of motion'.

This quantity of motion, which involves both mass and velocity, is called **momentum** and is quoted as the equation:

> **momentum = mass × velocity**
> ($kg\,m\,s^{-1}$)   (kg)   ($m\,s^{-1}$)

Mathematically expressed:

$$\underline{p} = m\,\underline{v}$$

Momentum is a **vector quantity**, and so problems involving momentum must be solved using vector mathematics that take the **magnitude** and **direction** into account.

When an unbalanced force acts on an object, it causes the momentum of the object to change. The size of the change in momentum (Δp) is calculated as:

> **change in momentum = final momentum – initial momentum**
> ($kg\,m\,s^{-1}$)   ($kg\,m\,s^{-1}$)   ($kg\,m\,s^{-1}$)

Expressed mathematically:

$$\Delta\underline{p} = \underline{p}_f - \underline{p}_i$$

The size of the change in momentum depends on the size of the **unbalanced force** AND the **time** for which the unbalanced force acts on the object. This is more commonly quoted as the equation:

> **net force = $\dfrac{\text{change in momentum}}{\text{time taken}}$** $(kg\,m\,s^{-1})$
> (N)   (s)

Expressed mathematically:

$$\underline{F}_{net} = \frac{\Delta\underline{p}}{\Delta t}$$

ISBN: 9780170195997

Momentum and collisions

The concept of **impulse** is usually used when a force acts for a short amount of time, such as during a collision. Impulse is the same as the change in the momentum of an object and can be expressed as:

change in momentum = force × time taken
$(kg\,m\,s^{-1})$ (N)   (s)

Mathematically expressed:

$$\Delta p = \underline{F}_{net}\,\Delta t$$

During any impact lasting time (t), the force rapidly increases to a maximum value and then falls back to zero. Consequently, questions involving impulse often refer to the **average force**.

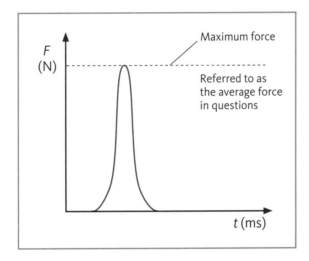

Exercise 4C

1   A cricket ball of mass 0.164 kg is travelling at 15.0 m s⁻¹ (left) towards a batsman. Calculate the momentum of the ball before it reaches the batsman.

_____

_____

_____

2   A tennis ball of mass 54.0 g has a momentum of 0.70 kg m s⁻¹ (right). Calculate the velocity of the ball.

_____

_____

_____

_____

ISBN: 9780170195997

3   A hockey player hits a stationary ball of mass 0.160 kg, causing it to move with a speed of 28 m s^{-1}.

  a   Given that the impact between the stick and the ball lasted 2.0 × 10^{-3} s, calculate the size of the average force applied to the ball.

_____

_____

_____

  b   Explain why this is an average force.

_____

_____

_____

4   Solve the following problem using $p = m\underline{v}$, $\Delta p = p_f - p_i$, and $\Delta\underline{p} = \underline{F}\Delta t$.

Initially a squash ball of mass 24.0 g is travelling at 28.0 m s^{-1} towards the wall. After the collision with the wall, the squash ball moves away in the opposite direction at 16.0 m s^{-1}.

  a   Calculate the change in momentum of the squash ball.

_____

_____

_____

_____

_____

_____

  b   Calculate the size of the force on the squash ball given that the impact takes 0.006 s.

_____

_____

_____

_____

5   Explain in terms of momentum why it is more difficult to stop a car full of passengers compared to a car with just the driver in it, even though both cars are travelling at the same speed.

_____

_____

_____

_____

_____

_____

_____

_____

_____

ISBN: 9780170195997

6 An impact attenuator is a road safety device designed to reduce the damage done to structures, vehicles and motorists that may result from a vehicle collision. The attenuator gradually collapses during a collision, thereby reducing the amount of damage done to the vehicle and the structure it is protecting. Explain how impact attenuators reduce the amount of damage.

Conservation of momentum

Principle of conservation of momentum

The principle of conservation of momentum states that the total momentum of a system of interacting (colliding or exploding) objects remains constant, provided there is no external net force acting on the system. This mathematical statement tells us that:

> TOTAL initial momentum of the system = TOTAL final momentum of the system
> $(kg\,m\,s^{-1})$ $(kg\,m\,s^{-1})$

Internal and external forces

According to Newton's Second Law, the momentum of an object will remain constant unless a **net external** force acts upon the object.

External forces can be identified as those forces that act on an object or set of objects (a system) from the **outside**, for example the force due to gravity.

Internal forces are the forces that act **between** the objects in a system (Newton's Third Law).

ISBN: 9780170195997

Worked example: Puck head to head with double puck

A single air hockey puck of mass 42.0 g slides at 3.00 m s^{-1} towards two pucks, one on top of the other, travelling in the opposite direction at 0.900 m s^{-1}. They collide head on and the double puck moves away at 1.57 m s^{-1} in a straight line as shown below. Calculate the final velocity of the single puck.

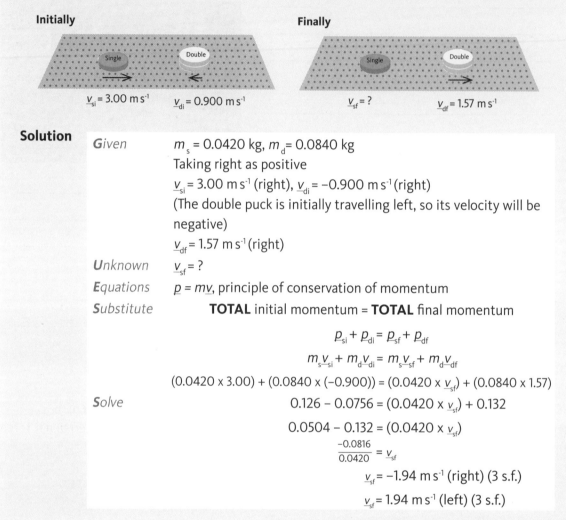

Initially

$\underline{v}_{si}$ = 3.00 m s^{-1} $\underline{v}_{di}$ = 0.900 m s^{-1}

Finally

$\underline{v}_{sf}$ = ? $\underline{v}_{df}$ = 1.57 m s^{-1}

Solution

Given

m_s = 0.0420 kg, m_d = 0.0840 kg
Taking right as positive
$\underline{v}_{si}$ = 3.00 m s^{-1} (right), $\underline{v}_{di}$ = −0.900 m s^{-1} (right)
(The double puck is initially travelling left, so its velocity will be negative)
$\underline{v}_{df}$ = 1.57 m s^{-1} (right)

Unknown $\underline{v}_{sf}$ = ?

Equations $\underline{p} = m\underline{v}$, principle of conservation of momentum

Substitute **TOTAL** initial momentum = **TOTAL** final momentum

$$p_{si} + p_{di} = p_{sf} + p_{df}$$

$$m_s\underline{v}_{si} + m_d\underline{v}_{di} = m_s\underline{v}_{sf} + m_d\underline{v}_{df}$$

$$(0.0420 \times 3.00) + (0.0840 \times (-0.900)) = (0.0420 \times \underline{v}_{sf}) + (0.0840 \times 1.57)$$

Solve

$$0.126 - 0.0756 = (0.0420 \times \underline{v}_{sf}) + 0.132$$

$$0.0504 - 0.132 = (0.0420 \times \underline{v}_{sf})$$

$$\frac{-0.0816}{0.0420} = \underline{v}_{sf}$$

$$\underline{v}_{sf} = -1.94 \text{ m s}^{-1} \text{ (right) (3 s.f.)}$$

$$\underline{v}_{sf} = 1.94 \text{ m s}^{-1} \text{ (left) (3 s.f.)}$$

Exercise 4D

1   On the following diagrams draw and label internal forces in red and external forces in blue.

   a   Inline skaters pushing apart.

ISBN: 9780170195997

b Dodgem cars bumping together just after the power is switched off.

c A tug-of-war.

2 A stack of three air hockey pucks of total mass 120.0 g slides at 2.40 m s^{-1} towards a stack of two pucks of mass 80.0 g, travelling in the opposite direction at 5.60 m s^{-1}. They collide head-on and the double puck moves away at 4.00 m s^{-1} in the opposite direction, as shown below.

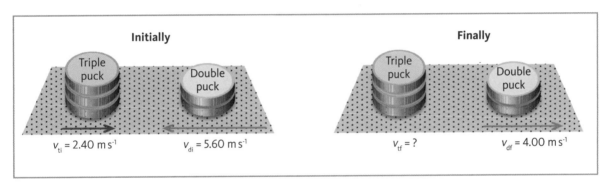

a State the principle of conservation of momentum and explain any assumptions you must make to solve this problem using the principle.

_____

_____

_____

_____

_____

b Calculate the initial momentum of the system.

_____

_____

_____

ISBN: 9780170195997

c   State the final momentum of the system, and hence determine the final velocity of the stack of three pucks.

_____

_____

_____

_____

_____

3   Hovercraft A (of mass 2.00×10^2 kg) is travelling at 10.0 m s⁻¹ when it bumps into the back of Hovercraft B (of mass 1.00×10^2 kg), which is moving at 7.0 m s⁻¹ in the same direction. After the collision Hovercraft A continues moving in the same direction but at a new speed of 8.0 m s⁻¹.

Initially		Finally	
Hovercraft A	Hovercraft B	Hovercraft A	Hovercraft B
10.0 m s⁻¹	7.0 m s⁻¹	8.0 m s⁻¹	$v_f = ?$

a   Calculate the change in momentum of Hovercraft A. Give your answer with an S.I. unit.

_____

_____

_____

Change in momentum:

                                        unit

b   State an alternative unit for momentum.

_____

c   State the change in momentum of Hovercraft B and explain how you came to your answer. Include any assumptions that you made in reaching your answer.

_____

_____

_____

_____

_____

d   Determine the final velocity of Hovercraft B.

_____

_____

_____

ISBN: 9780170195997

4 An empty red wagon of mass 1.00 tonne is rolling at a steady speed of 6.0 m s⁻¹ along a level track when it collides with a green wagon of mass 1.00 tonne, which is fully loaded with 9.00 tonnes of coal. The two wagons link together and continue moving in the same direction as the red wagon (1 tonne = 1000 kg).

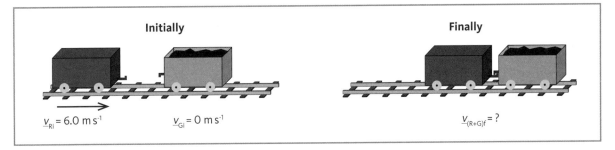

Initially **Finally**

$v_{Ri} = 6.0$ m s⁻¹ $v_{Gi} = 0$ m s⁻¹ $v_{(R+G)f} = ?$

a Calculate the initial momentum of the system.

b State the final momentum of the system, and hence determine the final velocity of the coupled (joined) wagons.

c Calculate the loss in momentum of the red wagon.

d Explain what happens to the momentum lost by the red wagon.

e The wagons continue to roll along the track joined together. When they reach the drop off point, the bottom of the green wagon opens and all the coal falls out as the wagons are still rolling. Describe what will happen to the momentum of the wagons, and explain how this will affect the speed at which they are rolling.

ISBN: 9780170195997

f An empty truck is rolling along a straight level track at a constant speed when it starts to rain very heavily straight down into the truck. Explain what will happen to the speed of the truck as it gradually fills with water.

_____

_____

_____

_____

_____

5 A water rocket (of mass 0.1 kg) is filled with 1.5 kg of water and is sitting horizontally at rest on a launching platform. When the launching clamp is released all the water is immediately sprayed backwards, travelling horizontally at a speed of 2.0 m s^{-1}.

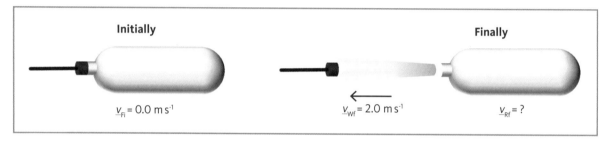

Initially

Finally

$v_{Fi} = 0.0$ m s^{-1}

$v_{Wf} = 2.0$ m s^{-1}

$v_{Rf} = ?$

a Calculate the velocity at which the rocket leaves the launcher.

_____

_____

_____

_____

b Initially neither the water nor the rocket has any momentum, but they both have momentum after the rocket is released. Does this contradict the principle of conservation of momentum? Explain your decision.

_____

_____

_____

_____

_____

_____

_____

ISBN: 9780170195997

4.5 Motion

Scalars: Distance and speed

Distance (*d*) is a scalar quantity and has size only. It can be defined as how far an object travels during a journey.

The speed (*v*) of an object is defined as the rate of change of distance, and can be written as:

speed	=	change in **distance**	(m)
(m s^{-1})		change in **time**	(s)

Expressed mathematically:

$$v = \frac{\Delta d}{\Delta t}$$

Vectors: Displacement, velocity and acceleration

Displacement (*d*) is a vector quantity and has size as well as a direction. It is defined as the straight line distance from a point in a particular direction.

Velocity (*v*) is a vector quantity and can be defined as the rate of change of displacement. It is written as:

velocity	=	change in **displacement**	(m)
(m s^{-1})		change in **time**	(s)

Expressed mathematically:

$$\underline{v} = \frac{\Delta \underline{d}}{\Delta t}$$

Acceleration (*a*) is also a vector quantity, and is defined as the rate of change of velocity (i.e. change in speed or direction or both). It can be determined by:

acceleration	=	change in **velocity**	(m s^{-1})
(m s^{-2})		change in **time**	(s)

Expressed mathematically:

$$\underline{a} = \frac{\Delta \underline{v}}{\Delta t}$$

Equations of motion (kinematic equations)

For situations where there is constant acceleration, the change in velocity is uniform and takes place in a straight line, these basic equations can be combined to produce five **equations of motion** (otherwise known as kinematic equations) that can be used to accurately describe the movement of an object.

To remember the equations of motion we can use the **Moving Hand** and **GUESS** techniques. Each term is assigned to a digit on your hand to remind you to check what terms have been **G**iven by the question, and what term is the **U**nknown. Each equation of motion contains four terms and has one term missing. Ticking off each term on your hand makes choosing the correct **E**quation from the formula list very straightforward. The values can then be **S**ubstituted and the equation **S**olved.

EQ	Equation	Used to solve problems involving ...	Missing term
1	$\underline{d} = \left(\dfrac{\underline{v}_i + \underline{v}_f}{2}\right)t$	$\underline{d}, \underline{v}_i, \underline{v}_f, t$	$\underline{a}$
2	$\underline{v}_f = \underline{v}_i + \underline{a}t$	$\underline{a}, \underline{v}_i, \underline{v}_f, t$	$\underline{d}$
3	$\underline{d} = \underline{v}_i t + \dfrac{1}{2}\underline{a}t^2$	$\underline{a}, \underline{d}, \underline{v}_i, t$	$\underline{v}_f$
4	$\underline{v}_f^2 = \underline{v}_i^2 + 2\underline{a}\underline{d}$	$\underline{a}, \underline{d}, \underline{v}_i, \underline{v}_f$	t
5	$\underline{d} = \underline{v}_f t - \dfrac{1}{2}\underline{a}t^2$	$\underline{a}, \underline{d}, \underline{v}_f, t$	$\underline{v}_i$

Moving Hand

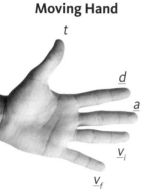

*T*ime is assigned to the **T**humb because it is the **only** scalar quantity. Each finger represents a vector quantity.

Be careful, because some quantities can be hidden in the wording. For example:
* an object may be described as being 'stationary' or 'at rest'. Both these statements mean it has zero velocity.
* an object travelling at a steady speed will have a zero acceleration.

Worked example: Moving on

The stopping distance for a car travelling 100 km h^{-1} (27.8 m s^{-1}) in a straight line is 78 m. Calculate the acceleration of the car while braking.

Solution

*G*iven

Taking FORWARD as positive
$\underline{v}_i = 27.8$ m s^{-1} (forward)
$\underline{v}_f = 0$ m s^{-1}
$\underline{d} = 78$ m (forward)

*U*nknown $\quad \underline{a} = ?$

*E*quations $\quad \underline{v}_f^2 = \underline{v}_i^2 + 2\underline{a}\underline{d}$

*S*ubstitute $\quad 0^2 = 27.8^2 + 2 \times \underline{a} \times 78$

*S*olve $\quad 0 = 772.84 + 156\underline{a}$

$-156\underline{a} = 772.84$

$\underline{a} = \dfrac{772.84}{-156} = -4.95$ m s^{-2}

$\underline{a} = -5.0$ m s^{-2} (forward) (2 s.f.)

ISBN: 9780170195997

Exercise 4E

1   A bird flies 270 m in a straight line in 2.0 minutes. Calculate the average speed of the bird and give your answer with an appropriate SI unit.

_____

_____

2   Caroline swims at a speed of 2.70 m s^{-1} over a distance of 71.54 m. Determine the time taken to cover the distance and give your answer to the correct number of significant figures.

_____

_____

3   An aircraft takes 3.0 hours to fly from Auckland to Melbourne travelling at an average velocity of 892 km h^{-1} (west).

   a   Calculate the displacement of the aircraft at the end of its journey. Give your answer in m.

_____

_____

   b   Explain why the aircraft is described as having an 'average velocity'.

_____

_____

_____

   c   The aircraft then returns by the same route but takes only 2.8 hours. Determine the average velocity of the round trip.

_____

_____

_____

4   A boat takes 12.0 hours to sail 252 km from Wellington to Nelson, as shown in the diagram at right. Nelson is displaced 122 km (west) of Wellington.

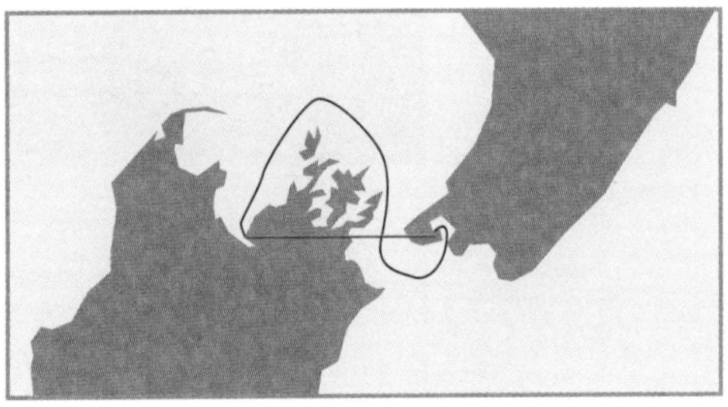

   a   Calculate the average speed.

_____

   b   Calculate the average velocity of the boat in km h^{-1}.

_____

_____

5 A car accelerates for 8.0 s at a steady rate from an initial velocity of 8.0 m s^{-1} (N) up to a final velocity of 13.8 m s^{-1} (N).

    a Calculate the average velocity of the car.

    _____

    _____

    b Calculate the displacement of the car.

    _____

    _____

    c Calculate the change in velocity of the car.

    _____

    _____

    d Calculate the acceleration of the car.

    _____

    _____

6 A motorbike travelling along a state highway at 28.0 m s^{-1} (N) approaches road works requiring the bike to slow to 8.33 m s^{-1} (N) in 6.0 s. Calculate the acceleration (size and direction) of the bike.

_____

_____

7 Monique is a goalkeeper practising saving shots. A soccer ball is kicked towards her at 16.7 m s^{-1}, which she punches away in the opposite direction with her hands. The collision with her hands takes 0.015 s, and the soccer ball moves away in the opposite direction at 12.2 m s^{-1}.

Initially	Finally
$\underline{v}_i = 16.7$ m s^{-1}	$\underline{v}_f = 12.2$ m s^{-1}

    a Calculate the average acceleration of the soccer ball.

    _____

    _____

ISBN: 9780170195997

b Explain why it is an average acceleration.

_____

_____

_____

_____

8 During a race, Akshay accelerates from rest to a velocity of 8.0 m s^{-1} (E) in 9.0 s.

a Calculate Akshay's acceleration.

_____

_____

_____

b Calculate the distance travelled while accelerating.

_____

_____

_____

9 The stopping distance, for a car travelling 60.0 km h^{-1} (16.6 m s^{-1}) in a straight line, is 36 m.

a Calculate the acceleration of the car while braking.

_____

_____

_____

b Calculate the time taken to bring the car to a stop.

_____

_____

_____

c Explain how doubling the initial speed of the car will affect the distance travelled before the car comes to rest and the time taken, assuming that it has the same acceleration.

_____

_____

_____

_____

ISBN: 9780170195997

Motion under gravity

On Earth, if we ignore the effects of air resistance, every falling object regardless of its mass experiences a uniform acceleration of $g = 9.8 \text{ m s}^{-2}$ downwards.

We can use the equations of motion to describe the displacement, velocity or time an object spends in the air as the acceleration throughout the flight is constant and downwards.

Important ideas to be aware of when solving vertical motion problems

- Choose **which direction is going to be positive** and state it.
  - Either make the initial direction of motion positive, or
  - Always make UP positive (similar to the positive direction on a graph).

- At the highest point in the flight (maximum displacement, d_{top}):
  - The **vertical velocity is zero**. The ball is still accelerating because it is changing direction.
  - The time taken to reach the top (t_{top}) is **the same as** the time taken to come back down if it lands at the same height at which it was released, in other words:

  $$t_{flight} = 2 \times t_{top}$$

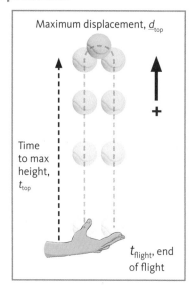

- When an object reaches the **same height** at which it was released:
  - The vertical **displacement will be zero**:

  $$d_f = d_i = 0$$

  - The final velocity will be the same size as the initial velocity but will be in the opposite direction, thus:

  $$v_f = -v_i$$

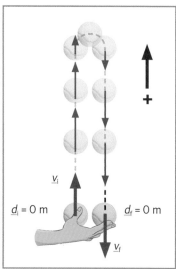

Projectile motion

When a ball is projected into the air at an angle to the vertical it will move sideways at the same time as it travels up and down. If friction is negligible, then during the flight the only force acting on the ball is due to gravity, and the ball will follow a **parabolic path**.

The acceleration due to gravity means that the ball's instantaneous velocity is constantly changing, so to understand the ball's motion we must resolve the initial velocity into the vertical and horizontal components.

ISBN: 9780170195997

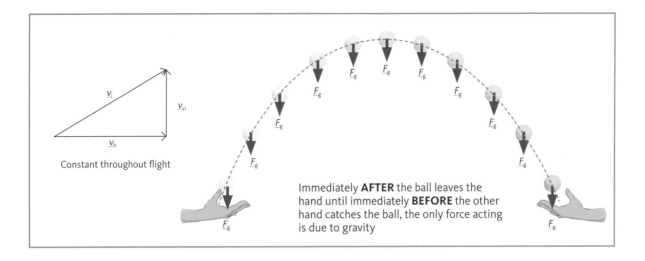

Constant throughout flight

Immediately **AFTER** the ball leaves the hand until immediately **BEFORE** the other hand catches the ball, the only force acting is due to gravity

Exercise 4F

When dealing with vertical motion on Earth, take the acceleration of gravity as g = 9.8 m s⁻².

1   The Mars explorer *Curiosity* landed on the surface of Mars on 5 August 2012. During the first stages of the landing, *Curiosity* fell into the Martian atmosphere before opening its parachute. Once the parachute was open, the spacecraft slowed from a speed of 405 m s⁻¹ at an altitude of 11 000 m to a speed of 125 m s⁻¹ at an altitude of 8000 m. The heat shield then separated from *Curiosity* at an altitude of 8000 m and fell towards the Martian surface.

   a   Determine the distance that *Curiosity* falls from 11 000 m to 8000 m.

   _____

   _____

   b   Calculate *Curiosity*'s average acceleration during this section of the journey.

   _____

   _____

   _____

   c   Calculate the velocity at which the heat shield hits the surface, given that the acceleration (*a*) due to gravity on Mars is 3.71 m s⁻². Assume air resistance on the heat shield can be ignored.

   _____

   _____

   _____

ISBN: 9780170195997

2   Alex throws a golf ball vertically upwards at a velocity of 8.4 m s⁻¹.

    a   Calculate the maximum height reached by the ball, and explain why the actual maximum height will be slightly less than your calculated value.

    _____

    _____

    _____

    _____

    _____

    _____

    b   Calculate the time taken for the ball to come back down to Alex's hand.

    _____

    _____

    _____

    c   Calculate the two times that the ball is travelling with a speed of 3.5 m s⁻¹.

    _____

    _____

    _____

3   Anna hits a hockey ball giving it an initial velocity of 26 m s⁻¹ at 40° to the horizontal.

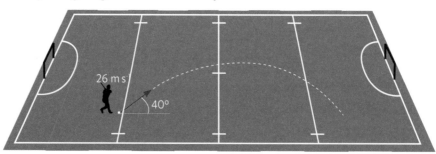

    a   Resolve the vector to determine the horizontal component of the velocity.

    _____

    b   Resolve the vector to determine the vertical component of the velocity.

    _____

    c   Calculate the maximum height the ball reaches during its flight.

    _____

    _____

    _____

    _____

    d   Determine the flight time of the ball.

    _____

    _____

    _____

ISBN: 9780170195997

e Calculate the maximum horizontal distance the ball travels before it hits the ground.

_____

_____

4 An aircraft is travelling horizontally when a skydiver jumps from the back door. The skydiver falls for 6.19 s and reaches a speed of 70.0 m s^{-1} at an angle of 30° to the vertical before opening the parachute. Assume air resistance is negligible prior to the parachute opening.

a Calculate the skydiver's vertical speed just before the parachute is opened.

_____

_____

_____

b Calculate the vertical distance the skydiver has fallen.

_____

_____

_____

c Calculate the initial velocity of the skydiver.

_____

_____

_____

d The skydiver falls like a projectile until the moment the parachute is opened. Explain why the skydiver no longer moves like a projectile once the parachute opens.

_____

_____

_____

5 Valerie is a shot putter who throws a shot put with a velocity of 13.64 m s^{-1} at an angle of 45.0° to the horizontal. The shot put leaves her hand at a height of 2.00 m above the ground and hits the ground 2.157 s later.

a Resolve the vector to determine the horizontal component of the velocity.

_____

b Resolve the vector to determine the vertical component of the velocity.

_____

ISBN: 9780170195997

c Calculate the horizontal displacement the shot put travels while it is in the air.

_____

_____

d Calculate the velocity (magnitude and direction) with which the shot put hits the ground.

_____

_____

_____

_____

e Calculate the time taken for the shot put to reach its maximum height.

_____

_____

_____

_____

f A friend suggests that Valerie might make the shot put go further if she throws it at the same speed but at an angle of 30° to the horizontal, as this will cause the shot put to travel faster horizontally. Explain whether the friend's advice is correct.

_____

_____

_____

_____

_____

_____

6 Olivia carries out an experiment to determine the time taken for a free falling object to travel different distances. She suspends a small ball bearing in a vacuum tube and uses an electromagnet to release the ball from rest when she starts the clock. She stops the clock when it passes different distance marks. Her data is shown in the following table.

Distance (m)	Time (s)				Processed data
	Trial 1	Trial 2	Trial 3	Average time (s)	
0.10	0.16	0.12	0.14		
0.20	0.21	0.20	0.19		
0.30	0.30	0.24	0.25		
0.40	0.29	0.28	0.30		
0.50	0.31	0.32	0.33		
0.60	0.35	0.36	0.34		

a Study the data and circle any anomalous points in the table.

ISBN: 9780170195997

b Explain why it is important that Olivia carries out the experiment three times and then calculates the average.

_____

_____

_____

_____

c Calculate the average time for each distance, and plot a graph of distance against average time.

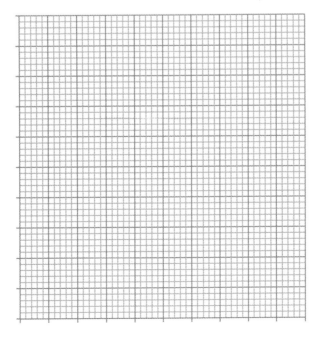

d State the relationship between distance and time based upon the shape of the graph.

_____

_____

e Process the data to produce a linear relationship and plot the graph.

ISBN: 9780170195997

f Determine the gradient of the line and the intercept, and state the equation of the line in the form of **y = mx + c.**

_____

_____

_____

_____

_____

g The relationship between distance (*d*) and time (*t*) for an object moving with a constant acceleration is given by the formula:

$$\underline{d} = \tfrac{1}{2}at^2 + \underline{v}_i t$$

i Comparing the equation of the line in **f** to the equation above, what does the intercept of the graph tell you about the motion of the ball?

_____

_____

_____

ii Calculate the acceleration of the ball.

_____

_____

_____

_____

_____

4.6 Forces and rotational motion

Torque (*τ*), also referred to as the moment of a force or leverage, refers to the turning effect caused by a force acting at a distance from a pivot, hinge or fulcrum.

Torque is defined as the turning effect due to a force (*F*) acting at a perpendicular distance (*d*) from a pivot. This is expressed as a word equation:

torque	**=**	**force**	**×**	**perpendicular distance**
(N m)		(N)		(m)

Expressed mathematically:

$$\tau = \underline{F}d_\perp$$

An unbalanced torque will cause an object to rotate in either a clockwise ($\tau_\circlearrowright$) or anticlockwise ($\tau_\circlearrowleft$) direction.

As the distance of the force from the pivot decreases, the torque decreases until the line of action of the force acts through the pivot, at which point the torque will be zero and there will be no turning effect on an object.

ISBN: 9780170195997

Equilibrium

If the sum of all the forces acting on an object is zero (ΣF = 0 N), or in other words, if:

- all the vertical forces are balanced ($F_{up} = F_{down}$), and
- all the horizontal forces are balanced ($F_{left} = F_{right}$),

then we can describe the object as being in **translational equilibrium** and it will either:

- remain stationary, or
- keep moving at a steady speed in a straight line.

If the sum of all the torques on an object is zero ($\Sigma \tau$ = 0 N m), or in other words, if:

- all the torques are balanced ($\tau_{anticlockwise} \circlearrowleft = \tau_{clockwise} \circlearrowright$),

then we can describe the object as being in **rotational equilibrium**, and an object will either:

- remain stationary, or
- keep rotating at a steady speed.

These two **equilibrium statements** are used to determine the forces acting on an object, such as a bridge, and by engineers when designing structures.

Solution techniques

- When a diagram is provided, annotate it to show all the forces acting on the object.
- Where there are two possible pivots in a problem, choose the pivot that does not involve the value you are trying to find (i.e. if you want to know the force acting on A, then make B your pivot).
- Annotate the distance of each force from the chosen pivot.

Exercise 4G

Where required use g = 9.8 N kg⁻¹.

1   John is holding a teapot of mass 1.50 kg at arms length, a distance of 60.0 cm from his shoulder.

   a   Show that a torque of 8.82 N clockwise is acting on John due to the weight of the teapot.

   _____

   _____

   _____

   b   Determine the size of the torque that John's muscles must supply to keep the teapot stationary in the air, and explain your answer.

   _____

   _____

   _____

   _____

   _____

   _____

c To avoid an unfortunate leak, John's shoulder muscles supply the necessary force to keep the teapot stationary. Calculate the size of the force produced by his muscles if they act at a distance of 4.0 cm from his shoulder.

_____

_____

2 Jake and Toby are identical twins who play soccer. They decide to have a competition to find out who can hold their soccer bag in the air for the longest time. They stand back to back and hold the bags at chest height. Both bags have the same mass but Toby holds his bag at arms length, and Jake holds his bag close to his chest. Toby loses the competition and tells Jake that the competition wasn't fair because his bag was heavier. Explain whether Toby's bag is heavier and whether the competition is fair.

_____

_____

_____

_____

_____

_____

_____

_____

3 Vernon, of mass 42.0 kg, sits on a tree branch at a distance of 3.0 m from the tree, causing the branch to bend downwards at an angle of 10.0° to the horizontal. Calculate the size of the torque produced by Vernon.

_____

_____

_____

_____

4 Cynthia, of mass 54.0 kg, and Stan, of mass 76.0 kg, are decorating the science building corridor. To reach the top of the wall they stand on a uniform plank resting on two short ladders 0.50 m from each end. The plank has a mass of 20.0 kg and is 5.00 m long. Cynthia stands 1.50 m from one end and Stan stands 1.00 m from the other end.

ISBN: 9780170195997

a Complete the diagram below by drawing labelled arrows to show all the forces acting on the plank.

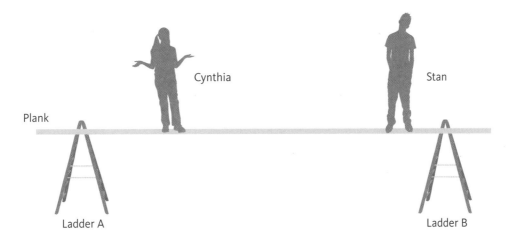

b Calculate the size of the support forces supplied by the two ladders.

_____

_____

_____

_____

_____

_____

_____

_____

_____

5 A window cleaner of mass 84.0 kg is standing on a platform lowered down from the top of a building by two wires, X and Y, attached at either end. The platform is 3.00 m long and has a mass of 135 kg. The cleaner stands 1.00 m from wire X and the bucket is 0.750 m from wire Y. Wire X supplies a support force of 1225 N.

a Complete the diagram below by drawing labelled arrows to show all the forces acting on the platform.

b Determine the size of the support force on wire Y and the mass of the window cleaner's bucket.

_____

_____

_____

_____

_____

_____

_____

_____

6 A glassblower holds a ball of glass of mass 0.80 kg on the end of a 1.20 m long pipe. The pipe has mass of 1.80 kg and is supported by the glassblower's left hand, 0.40 m from the blowing end. The glassblower's right hand holds the pipe stationary at the blowing end.

a Complete the diagram below by drawing labelled arrows to show all the forces acting on the pipe.

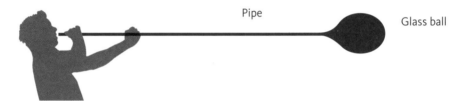

Pipe

Glass ball

b Determine the size of the support force supplied by the left hand and the balancing force supplied by the right hand.

_____

_____

_____

_____

_____

_____

_____

_____

ISBN: 9780170195997

c Explain why the glassblower leans back when he is holding the pipe and glass ball out in front of him.

_____

_____

_____

_____

_____

_____

7 The Gateshead Millennium Bridge is a special type of suspension bridge. Its lower arch has a footpath for pedestrians and cyclists and is supported by long wires attached to an upper arch.

Geordie builds a simplified model, as shown in the diagram below. On Geordie's model the lower arch is supported by a single cable, attached to the centre of the arch 63.0 cm from the pivot and at an angle of 40.0° to the horizontal. The tension force in the cable is 35.0 N. The bridge is in equilibrium.

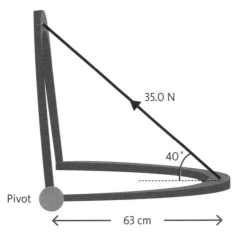

a Explain the meaning of the term equilibrium.

_____

_____

_____

_____

b Show that the vertical component of the tension force in the cable is 22.5 N.

_____

_____

ISBN: 9780170195997

c The lower arch has a weight of 31.50 N. By considering the torques on the lower arch, determine the position of the centre of mass of the lower arch and draw it on the diagram.

_____

_____

_____

_____

d The bridge can be opened when tall ships need to pass under it, by rotating it about the pivot. Discuss how the equilibrium of the bridge changes as the bridge starts to rotate, moves in a circle at a steady speed, and comes to rest in its open position.

_____

_____

_____

_____

_____

_____

_____

_____

_____

_____

4.7 Circular motion

Motion in a circle

Consider an object moving at a steady speed (v) around the circumference (C) of a circle of constant radius (r). The length of the path completed by the object during a single rotation or **revolution** is given by the formula:

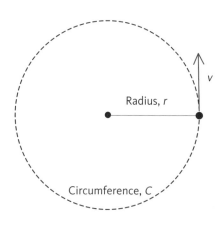

$$C = 2\pi r$$

The time taken to complete a single revolution is called a **period** (T). The speed of an object completing a single revolution of a circle can therefore be found using the formula $v = \dfrac{\Delta d}{\Delta t}$.

$$v = \frac{2\pi r}{T}$$

The number of times (n) an object completes a revolution each second is referred to as the **frequency** (f), and can be calculated using $f = \dfrac{n}{t}$. The unit of frequency is 'per second', expressed as the SI unit the **hertz** (Hz) where 1 Hz = 1 s⁻¹. Since one revolution is completed each period, the frequency is related to the time period, and so:

$$f = \frac{1}{T}$$

ISBN: 9780170195997

Centripetal acceleration

When an object changes its velocity, it experiences an acceleration in the direction of the change in velocity. So for an object to move along a circular path, it must be continually accelerating towards the centre of the circle. This is known as **centripetal acceleration** (a_c) and can be calculated using:

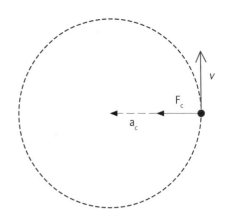

centripetal acceleration	=	speed²	(m s⁻¹)²
(m s⁻²)		radius	(m)

Expressed mathematically:

$$a_c = \frac{v^2}{r}$$

Centripetal force

According to Newton's Second Law, $F = ma$, ie. a **net force** (F_{net}) is required to cause an object of mass (m) to accelerate (a) in the direction of the force, hence:

$$F_c = \frac{mv^2}{r}$$

The centripetal force acts at right angles to the direction of motion, so it has no effect on the speed of the object but will cause it to change direction.

If the centripetal force is removed then the object will travel in a straight line at a tangent to the circle at a steady speed, unless acted on by other external forces such as gravity or friction, in agreement with Newton's First Law.

 Worked example: Car on ice

As the car is taking the bend it hits a patch of frictionless ice. Describe and explain what will happen to the motion of the car.

Solution

Given 'Frictionless ice' means that there will be no friction to provide the centripetal force, no net external force.

Unknown 'motion of the car' implies velocity and acceleration.

Equations $F_{net} = ma$ and $a = \frac{\Delta v}{\Delta t}$

Substitute My **FAV**o**U**ri**Te**: $\overrightarrow{F_{net}} = m\vec{a}$ so $\underline{a} = \frac{\Delta v}{\Delta t}$ 0
$\quad\quad\quad\quad\quad\quad 0 \quad = 0 \quad 0$

Solve As the ice is frictionless there will be no friction acting on the car to provide the centripetal **f**orce. In the absence of a net external force the car cannot **a**ccelerate, so there will be no change in the **v**elocity. The car will continue moving at a steady speed in a straight line, on a tangent to the point where the centripetal force disappeared.

ISBN: 9780170195997

Exercise 4H

1   A motorbike drives at a steady speed of 11 m s^{-1} around a roundabout of radius 24 m.

    a   Calculate the time taken by the motorbike to complete a single revolution.

_____

_____

    b   Calculate the centripetal acceleration of the motorbike.

_____

_____

    c   Explain why the motorbike can have an acceleration even though it is travelling at a steady speed.

_____

_____

_____

_____

_____

_____

_____

_____

2   Tahi swings a purerehua (bull roarer) on a long piece of string around his head at 144 rpm.

    a   Show that there are 2.4 revolutions per second and state an alternative unit.

_____

_____

Frequency _____ unit _____

    b   Calculate the time taken to complete a single revolution.

_____

_____

    c   The purerehua travels in a circle of diameter 2.2 m. Calculate the distance travelled by the purerehua during one revolution.

_____

_____

ISBN: 9780170195997

d Determine the size of the force acting on the purerehua if it has a mass of 280 g.

_____

_____

_____

_____

_____

_____

_____

e Explain what will happen to the purerehua if the string holding it suddenly snaps. In your answer you should consider the motion in both horizontal and vertical directions, and state any assumptions that you have made.

_____

_____

_____

_____

_____

_____

_____

_____

_____

_____

_____

3 A force of 7872 N acts on a car of mass 1025 kg, causing it to travel at a steady speed around a bend of radius 75.0 m.

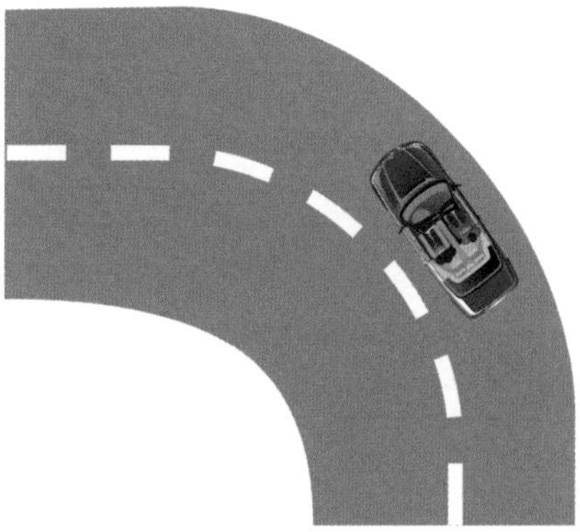

a Draw an arrow on the car in the diagram to show the direction of the force acting on the car, and explain the reasons for the direction in which your arrow points.

_____

_____

_____

_____

_____

_____

b   Calculate the speed at which the car is travelling.

_____

_____

_____

_____

_____

_____

c   A passenger in the car complains about feeling like she is being thrown outwards against the door of the car as it travels around the bend. Is her description of feeling thrown outwards correct? Explain your answer.

_____

_____

_____

_____

_____

_____

_____

_____

d   Calculate the size of the force of the passenger on the door of the car if she has a mass of 62 kg.

_____

_____

_____

_____

_____

e   Describe exactly what will happen to the force on the passenger if the driver doubles the amount of time it takes to drive around the bend.

_____

_____

_____

_____

_____

_____

_____

_____

ISBN: 9780170195997

4.8 Hooke's Law

When a **load force** (F_L) is applied to a spring, the length increases. The amount it increases is called the **extension** (x). A graph of the force against extension reveals a linear relationship up to a point called the **limit of proportionality**, beyond which the spring becomes permanently deformed.

Up to the limit of proportionality the gradient of the graph is a constant for each individual spring and is called the **spring constant** (k), measured in $N\,m^{-1}$ where:

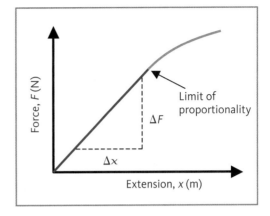

$$k = \frac{\Delta F}{\Delta x}$$

Rearranging this gives the formula:

$$F_L = kx$$

In equilibrium, the spring produces an equal but opposite force to the load force that is stretching (or compressing) it. This is called a **restoring force (F_r)** and acts in the opposite direction, so:

$$F_R = -kx$$

The negative sign indicates that the force from the spring acts in the opposite direction to the extension.

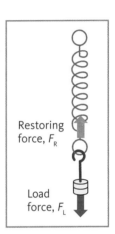

Parallel spring systems

When cables or springs are placed in parallel (side-by-side), they will each take a share of the load and the extension or compression of the each cable or spring is reduced. The result will be equivalent to a single spring with a greater spring constant.

Series spring systems

When cables or springs are placed in series (end-to-end), they both experience the same load and so the overall extension is increased. The result will be equivalent to a single spring with a lower spring constant.

Elastic potential energy

Work must be done to change the shape of an elastic object by stretching or compressing it. The work done by a force (F) stretching or compressing a spring with a spring constant (k) will cause an extension (x). This will result in **energy** (E_p) being stored in the spring.

ISBN: 9780170195997

The amount of energy stored in a stretched (or compressed) spring can be determined by finding the area (= ½ × height × base) under a graph of force against extension.

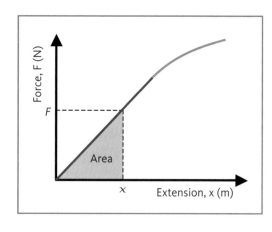

energy = ½ × force × extension
(J)   (N)   (m)

Expressed mathematically:

$$E_p = \frac{1}{2} Fx$$

If this is then combined with Hooke's Law ($F = kx$) it can be rewritten as:

$$E_p = \frac{1}{2} kx^2$$

Worked example: Energy stored in a spring

When Rebecca stands on a trampoline it stretches 6.0 cm downwards, but when Daniel stands on the trampoline it only stretches 3.0 cm. Discuss how this affects the energy stored in the trampoline.

Solution

Given
x_R = 6.0 cm = 0.06 m
x_D = 3.0 cm = 0.030 m
so $x_D = ½ x_R$

Unknown
E_p = ?

Equations
$E_p = ½ kx^2$

Substitute
$E_p = ½ kx^2$
¼ ↓ = ↓ (½)²

Solve
The spring constant of the trampoline is the same, so halving the extension will result in the elastic potential energy stored in the trampoline by Daniel, only being a quarter of the energy stored in the trampoline when Rebecca stands on it, $E_p = ½ kx^2$.

ISBN: 9780170195997

Exercise 4I

Where required take g = 9.8 N kg⁻¹.

1   The spring in a baby bouncer stretches 16 cm when a baby of mass 7.2 kg is sitting in it.

   a   Calculate the spring constant of the spring and state its unit.

   _____

   _____

   _____

   Spring constant _____ unit _____

   b   Calculate the energy stored in the spring.

   _____

   _____

   _____

2   Geordie has made a model of Gateshead Millennium Bridge. In his model the support cable is made of string and it experiences a tension force of 35.0 N. The spring constant of the string is 875 N m⁻¹.

   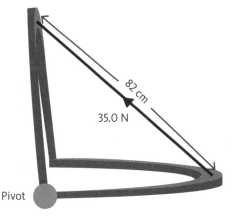

   a   Calculate the extension of the support cable.

   _____

   _____

   _____

   b   Calculate the original length of the support cable if the stretched cable is 82 cm long.

   _____

   _____

   _____

ISBN: 9780170195997

c Geordie then adds two more identical strings next to the original string to improve the strength of the support cable. What will be the new stretched length of the support cable?

_____

_____

_____

_____

_____

_____

d Explain how using the combined strings will affect the energy stored in the support cable.

_____

_____

_____

_____

_____

_____

_____

_____

3 Six children, each of mass 30.0 kg, sit on a children's playground ride causing the four springs supporting it to compress by 45.0 mm.

Platform

Springs

a Calculate the spring constant of a single spring.

_____

_____

_____

_____

b The children then start to bounce on the seats of the ride. Label the diagram (above) to show all the forces acting on the red platform when the ride is at the bottom of a bounce. (You may consider the combined weight of the children and ride as a single force.)

ISBN: 9780170195997

c Calculate the energy stored in each spring when the ride is at rest.

_____

_____

_____

_____

d The manufacturer of the ride believes that the ride will be more 'entertaining' if they add another set of identical springs on top, as shown in the diagram. Discuss how adding the extra springs will change the ride in terms of the compression of the springs and the energy stored in the springs.

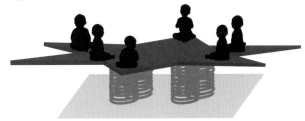

_____

_____

_____

_____

_____

_____

_____

4 Manoj and Shantelle build a model eardrum to investigate the effects of loud sounds on the human ear. They apply an increasing force to the model eardrum and measure the amount it deflects inwards. Their data is shown in the table below.

Force, F (N)	Deflection, x ($\times 10^{-3}$ m)			Average deflection ($\times 10^{-3}$ m)
	Trial 1	Trial 2	Trial 3	
15	0.04	0.03	0.03	
30	0.06	0.06	0.07	
45	0.10	0.08	0.09	
60	0.12	0.11	0.13	
75	0.15	0.14	0.15	
90	0.19	0.18	0.18	

a Complete the table by calculating the average deflection for each applied force.

b Plot a graph of force (y-axis) against average deflection (x-axis).

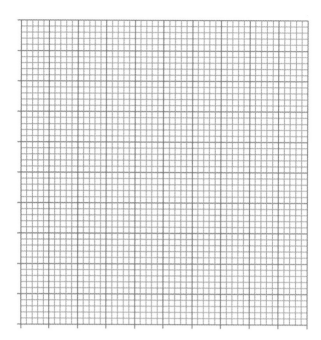

c State the relationship between force and deflection based upon the shape of your graph.

_____

_____

d Determine the gradient of the line, its unit and the value of the intercept.

_____

_____

_____

e State the equation of the line in the form of $y = mx + c$.

_____

f The relationship between force and the deflection of the eardrum is given by the formula $F = kx$.

By comparing the equation of the line in **e** to the equation above, determine the spring constant of the model eardrum.

_____

_____

_____

_____

_____

ISBN: 9780170195997

g Determine the area under the graph line and its unit. With reference to the area equation and the unit of the answer, explain what this quantity represents.

_____

_____

_____

_____

_____

_____

_____

4.9 Work, energy and power

Work

In science the word 'work' has a very specific meaning, which is different from the everyday use of the word. Work is defined as:

work done = force × distance moved by the force in the direction of the force
(N m)   (N)   (m)

Expressed mathematically:

$$W = F \times \underline{d}$$

If a force acts at an angle to the direction of motion, then the work done can be determined by resolving to find the component of the force in the direction of motion.

An object that is capable of doing work is said to possess energy, measured in **joules** (J). The amount of energy that an object possesses is equal to the amount of work done on it to give it that energy, which is the same as the amount of work the object could do. In other words:

work done   =   energy changed
(N m)   (J)

Expressed mathematically:

$$W = \Delta E$$

Energy

The **principle of conservation of energy** is one of the most important and fundamental laws in science. It states that:

Energy cannot be created or destroyed, but it can be transferred from one form to another.

There are many different types of energy, but they can be grouped into two forms: potential (stored) energy; and kinetic (active) energy.

ISBN: 9780170195997

Potential (stored) energy	Kinetic (active) energy
gravitational	kinetic (moving) energy
electric	heat
magnetic	sound
elastic	radiant
chemical	
nuclear	

Gravitational potential energy

Work is done against gravity (g) when an object of mass (m) is lifted through a vertical **distance** (Δh), causing the object to gain **gravitational potential energy** (ΔE_p) such that:

$$\Delta E_p = mg\Delta h$$

Or, using words:

change in gravitational energy = mass × acceleration due to gravity × change in height
(J) (kg) (m s⁻²) (m)

Kinetic energy

Kinetic energy (E_k) can be defined as the energy due to the motion of an object, and is equal to the work done accelerating the object to that speed or bringing the object to rest. It is given by the formula:

$$E_k = \frac{1}{2}m\underline{v}^2$$

Or, using words:

kinetic energy = ½ × mass × (velocity)²
(J) (kg) (m s⁻¹)²

Power

The power of a machine is the rate at which it does work, and can be expressed as:

$$\textbf{power} = \frac{\textbf{energy changed}}{\textbf{time taken}} \quad \begin{array}{l}(J)\\(s)\end{array}$$

(Js⁻¹)

Expressed mathematically:

$$P = \frac{\Delta E}{t}$$

The unit of power is the **watt** (W), where:

$$\textbf{1 W = 1 Js}^{-1}$$

ISBN: 9780170195997

Since power refers to the **changing** of energy rather than the energy stored, problems involving power will often refer to the 'active energies' rather than specifically asking about the power. The following 'power words' can help to identify a power question.

Active energy	kinetic	heat	sound	radiant
Examples of power words	Work rate Powerful	Hotter Coldest	Loudest Quieter	Brighter Dimmest

Exercise 4J

1   A woman pulls a suitcase 32.0 m across level floor with a force of 28 N at an angle of 30° to the horizontal.

  a   Resolve the force vector into horizontal and vertical components.

  _____

  _____

  _____

  _____

  b   Determine how much work the woman does moving the case.

  _____

  _____

  _____

  _____

  c   If the woman covers the distance in 48 s calculate her average power output.

  _____

  _____

  _____

  _____

ISBN: 9780170195997

2 Gareth, Ivan and Tom push a small car 25 m up a hill inclined at 19° to the horizontal. The car has a weight of 8600 N and they push it at a steady speed of 0.5 m s⁻¹ against an opposing force of friction of 70 N acting down the hill.

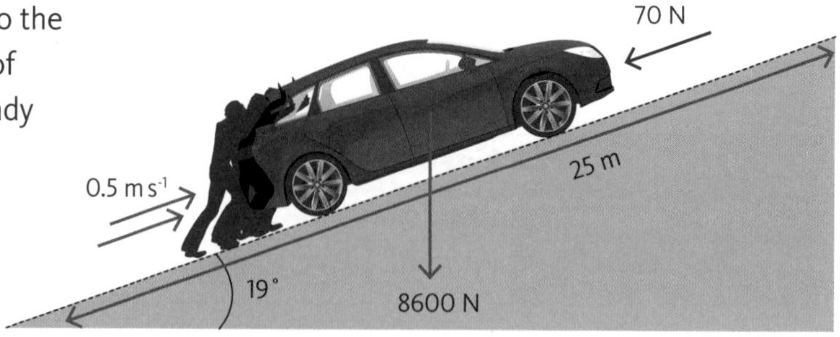

70 N

25 m

0.5 m s⁻¹

19°

8600 N

a By considering all the forces acting on the car, calculate the net force.

_____

_____

_____

b Determine the size of the boys' combined push force.

_____

_____

_____

_____

_____

c Determine the combined average power output of the boys.

_____

_____

_____

_____

3 An arrow of mass 0.020 kg is moving at 90.0 m s⁻¹ when it strikes a target and comes to rest.

a Show that the arrow has 81 J of kinetic energy just before it hits the target.

_____

_____

_____

_____

_____

_____

_____

ISBN: 9780170195997

b Determine the work done by friction bringing the arrow to a stop and explain your answer.

_____

_____

_____

_____

c The arrow travels 6.0 cm into the target before coming to rest. Calculate the average force provided by friction to stop the arrow.

_____

_____

_____

_____

4 A Porsche Carrera accelerates from 0 to 28 m s^{-1} in 4.8 s covering a distance of 67.2 m along a straight level track. (Assume the mass of the car and its driver together is 1530 kg.)

a Calculate the acceleration of the car.

_____

_____

_____

b Calculate the average force produced by the car.

_____

_____

c Calculate the average power output of the car during this time.

_____

_____

_____

_____

_____

d Explain why the average power converted by the engine is greater than the power output calculated in **c**.

_____

_____

_____

_____

_____

ISBN: 9780170195997

5   Kit and Mel carry out an investigation to find the relationship between the elastic potential energy stored in a spring and the extension of the spring. They compressed a spring and used it to launch a mass of 0.50 kg vertically upwards. The height gained by the mass could then be used to determine the energy in the spring. All their readings were accurate to 1 mm, and their data is shown below.

Compression of the spring (mm)	Height gained by the mass (mm)	Compression of the spring (. . .)	Height gained by the mass (. . .)	Change in energy (. . .)	Processed data
10	23				
20	92				
30	207				
40	367				
50	574				
60	827				

a   Convert the compression and height data to an appropriate SI unit and record it in the table to the correct number of significant figures.

b   Calculate the energy changed by the mass using the formula $\Delta E_p = mg\Delta h$ for each of the different heights, and record the energies in the table. (Taking $g = 9.8$ N kg^{-1}.)

c   Plot a graph of energy (y-axis) against compression (x-axis).

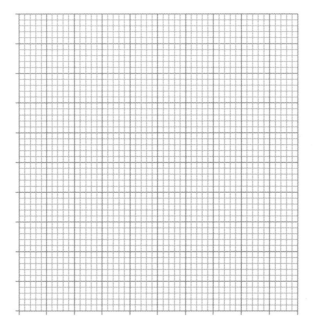

d   State the relationship between energy and compression based on the shape of your graph.

_____

_____

ISBN: 9780170195997

e Process the data to produce a linear relationship and plot the graph.

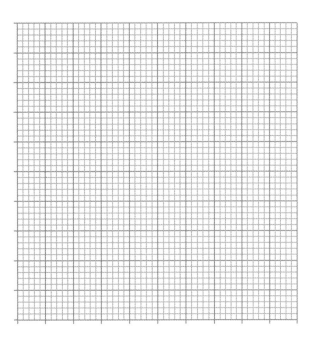

f Determine the gradient and intercept of the graph.

_____

_____

_____

g State the equation of the line in the form of $y = mx + c$.

_____

_____

h The relationship between energy and the extension of a spring is given by the formula $E_p = \dfrac{1}{2}kx^2$.

By comparing the equation of the line in **g** to the equation above, determine the spring constant of the spring.

_____

_____

_____

_____

i Explain how the energy due to the height of the mass can be used to determine the energy stored in the spring. State any assumptions that you are making.

_____

_____

_____

_____

_____

ISBN: 9780170195997

5 Atomic and nuclear physics

5.1 Models of the atom

Thomson's model of the atom

Thomson's 'plum pudding' model comprised of a sphere of positive charge with electrons embedded in it like 'raisins in a plum pudding'. The model could be used to explain:

- The chemical properties of different elements
- The electrical neutrality of an atom
- The source of electrons, which had only just been identified by Thomson as a subatomic particle.

But the plum pudding model had a number of weaknesses and could not explain:

- How the electrons were arranged in the positive sphere
- Why different elements give out different colours when heated (for example, neon street lights glow orange but nitrogen lamps glow purple)
- Why some elements release ionising radiation.

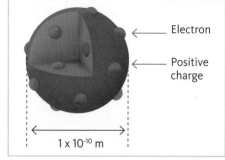

Rutherford's experiment

In 1910, Ernest Rutherford, Hans Geiger and Ernest Marsden designed an experiment to test Thomson's model of the structure of an atom, by bombarding a thin gold foil with alpha particles (positive helium ions). If Thomson's model were correct, all alpha particles should pass through the gold atoms without significant deviation.

The experiment involved the following apparatus:

- A radioactive source (radium) released alpha particles towards a lead screen

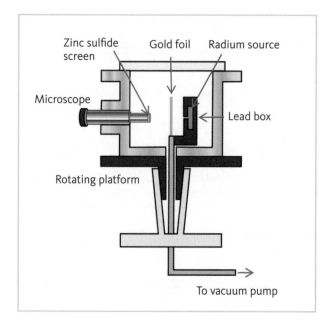

ISBN: 9780170195997

through an evacuated chamber, to prevent the alpha particles from being absorbed or deflected by air molecules.

- The alpha particles passed through holes in a lead screen to form a narrow beam. Any particles that hit the lead were absorbed.
- The beam of alpha ($\propto$) particles then hit a thin gold foil (a few 100 atoms thick).
- A zinc sulfide screen was placed in front of a microscope. When struck by a charged particle the screen would fluoresce.

The observations

Geiger and Marsden made three observations:

1 Some were scattered through large angles.
2 The majority of alpha particles passed straight through.
3 A very small number (1 in 8000) were deflected by more than 90° (they bounced back!).

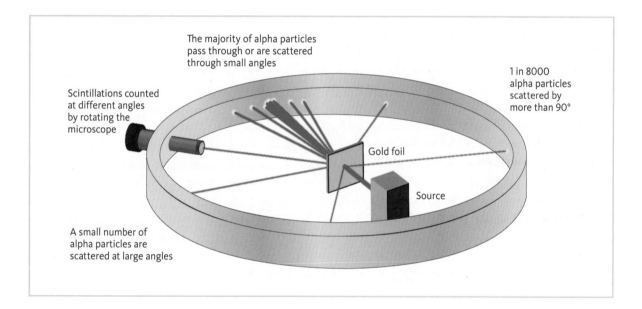

Rutherford considered each of the observations and drew certain conclusions:

Observation 1

Some alpha particles were scattered through large angles.

Conclusion

The alpha particles that were scattered through large angles must have encountered a **concentrated positive charge**.

Observation 2

The majority of alpha particles passed straight through.

Conclusion

If most of the alpha particles passed through the gold foil with little or no scattering, the chance of an alpha particle passing close to the positive charge must be very small. Given that the foil was several hundred atoms thick the positive charge must have been **significantly smaller than the atom**. Most of the atom must be empty space in which the electrons orbit at a great distance from the positive charge, so that the negative charge of the electrons did not act as a shield when an alpha particle entered the atom.

Observation 3

A very small number (1 in 8000) were deflected by more than 90° (they bounced back!).

Conclusion

The alpha particles that were scattered backwards must have encountered something with a much **larger mass** and concentrated positive charge.

Rutherford's model of the atom

- The positive charge is concentrated in a dense, positively charged nucleus that contains most of the mass of the atom.
- The nucleus was surrounded by orbiting electrons, held there by the electrostatic force of attraction between the negatively charged electrons and the positively charged nucleus.
- The total positive charge of the nucleus was equal to the negative charge of the electrons, so the atom was neutral overall.
- The electrons occupied most of the volume of the atom.

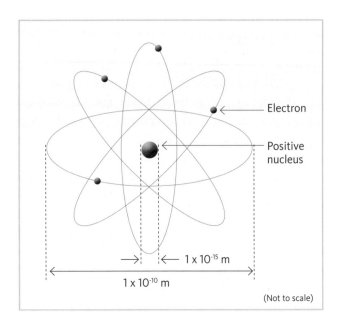

Electron

Positive nucleus

1×10^{-15} m

1×10^{-10} m

(Not to scale)

The Rutherford–Bohr model of the nucleus

In 1919 Rutherford discovered the **proton**, which had a positive charge the same size as an electron, but was nearly 2000 × more massive. Then, in 1932, **James Chadwick** discovered the **neutron**, a neutral particle.

The new model of the atom composed of neutrons and protons, referred to as **nucleons** as they are found in the nucleus of the atom, and electrons orbiting around the outside in allowed orbits.

Atoms have a neutral charge overall because there are an equal number of protons and electrons.

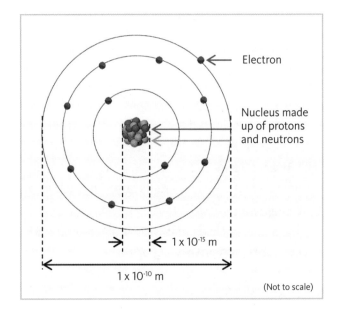

Electron

Nucleus made up of protons and neutrons

1×10^{-15} m

1×10^{-10} m

(Not to scale)

Ions

If enough energy is supplied to an atom an electron can be completely removed. This leaves the atom with a positive charge and is called a positive **ion**. Atoms can also gain extra electron and become negative ions.

ISBN: 9780170195997

Exercise 5A

1   In 1903 Thomson proposed a model of the atom based upon his discovery of the electron.

   a   Draw a labelled diagram of Thomson's model of the atom in the space provided, and explain why it was called the 'plum pudding' model.

   _____

   _____

   _____

   _____

   _____

   _____

   b   State the charge on Thomson's atom and explain how his model of the atom could explain the existence of different elements.

   _____

   _____

   _____

   c   Describe two ways in which Thomson's model failed to explain the behaviour of elements.

   _____

   _____

   _____

   _____

   _____

   _____

   _____

   _____

   _____

ISBN: 9780170195997

2 In 1911 Rutherford proposed a model of the atom based upon his gold foil experiments.

a Draw a labelled diagram of Rutherford's model of the atom in the space provided and explain why it was called the 'planetary' model.

b Describe two similarities between Thomson's model and Rutherford's model.

c Describe one important difference between Thomson's model and Rutherford's model.

d Describe how Rutherford's model is different to the Bohr model of the atom.

ISBN: 9780170195997

3 Rutherford carried out an experiment that involved firing alpha particles at a gold foil and observing what happened to them.

    a Label the following diagram of Rutherford's gold foil apparatus.

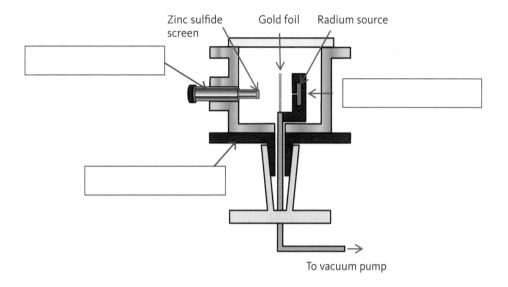

Zinc sulfide screen   Gold foil   Radium source

To vacuum pump

    b Explain the purpose of the zinc sulfide screen.

_____

_____

_____

    c Why did the experiment have to be carried out in a vacuum?

_____

_____

_____

_____

    Rutherford made three important observations that helped him develop his model of the atom. The diagram below represents his findings.

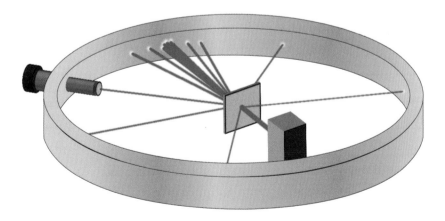

ISBN: 9780170195997

d State the three observations Rutherford made during his experiment.

i _____

_____

ii _____

_____

iii _____

_____

e State and explain Rutherford's conclusions based on each observation.

_____

_____

_____

_____

_____

_____

_____

_____

_____

_____

_____

_____

_____

_____

_____

_____

_____

4 In 1911 Rutherford carried out an experiment to test Thomson's model of the atom. He fired alpha particles at a gold foil (only a few 100 atoms thick), and then observed what happened to the alpha particles. The experiment was carried out in a vacuum because alpha particles will ionise air molecules and quickly lose energy.

a Explain what is meant by the term ionise.

_____

_____

_____

_____

ISBN: 9780170195997

b Describe and explain what Rutherford would have observed if Thomson's model of the atom was correct.

_____

_____

_____

_____

_____

_____

c Describe the observations that Rutherford made that supported Thomson's model.

_____

_____

_____

d Describe the observations that Rutherford made that contradicted Thomson's model.

_____

_____

_____

_____

_____

_____

e Describe Rutherford's model of the atom and explain how it overcomes the failings in Thomson's model.

_____

_____

_____

_____

_____

_____

_____

_____

_____

_____

_____

_____

ISBN: 9780170195997

5.2 Radioactivity

Proton and nucleon numbers

- **Proton number** (Z) (or **atomic number**) = the number of protons.
- **Nucleon number** (A) (or **mass number**) = the number of protons **+** neutrons.

To find the number of neutrons in a nucleus:

$$N = A - Z$$

The standard notation for representing a particular atom (nuclide) of an element is shown below.

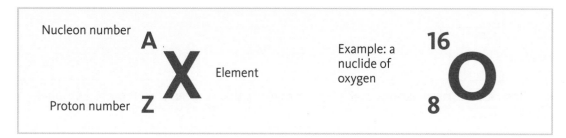

Any sample of a naturally occurring element will contain several different nuclides – these are collectively referred to as **isotopes**. Some of these nuclides are stable and do not change, however some nuclides are **unstable** and change into different nuclides, emitting alpha, beta and/or gamma radiation in the process.

Unstable nuclides: Radionuclides

There are a number of reasons why a nuclide might be unstable:

1 **Too few neutrons**. The protons in a nucleus all have a positive charge so are repelling each other, however a **very short range** force called the **strong force** acts between all the particles in the nucleus and is 100 times stronger than the electrostatic force. The neutrons provide the extra 'glue' necessary to hold the nucleus together.

2 **Too many neutrons**. Neutrons are not stable on their own and will decay in about 15 minutes. Inside a nucleus the interaction between the protons and the neutrons maintains the neutrons stability.

3 **Too much energy**. If the nucleus has too much potential and kinetic energy.

When an unstable nuclide (called the **parent nuclide**) emits radiation it loses energy and transforms or **decays** into a new element (called the **daughter nuclide**) with a more stable proton-neutron ratio. The daughter nuclide and the emitted particle are referred to as the **decay products**.

Nuclear conservation laws

During a radioactive decay the **proton number**, **nucleon number**, **charge** and **mass-energy** are conserved, hence:

sum of the nucleon numbers before the reaction = sum of the nucleon numbers after the reaction

sum of the proton numbers before the reaction = sum of the proton numbers after the reaction

ISBN: 9780170195997

Ionising radiation

In Level 2 Physics, three types of ionising radiation are considered – alpha, beta and gamma – as shown in the table below.

	Alpha radiation	Beta radiation	Gamma radiation
Radiation			
Symbol	$^4_2\alpha$, ^4_2He or as $^4_2He^{2+}$	$^0_{-1}\beta$ or as $^0_{-1}e$	$^0_0\gamma$
Description	Made up of two protons and two neutrons, but no electrons, the alpha particle is like a high speed **helium ion** emitted from the nucleus.	Beta particles are high speed electrons, formed when a **neutron turns** into a **proton** and an **electron**. As they are emitted from the nucleus they are often called nuclear electrons.	Gamma radiation is part of the **electromagnetic spectrum** so has no mass and no charge.
Speed	About 5% of the speed of light (0.15×10^8 m s⁻¹).	Have a **range** of speeds up to 99% the speed of light (2.97×10^8 m s⁻¹).	They travel at the speed of light (3.00×10^8 m s⁻¹).
Ionising ability	Due to their relatively large mass and charge, alpha particles have a **strong ionising effect**.	Beta particles have half the charge and $\frac{1}{8000}$ of the mass of an alpha so have a much **weaker ionising effect**.	Gamma radiation has no charge so it has an **extremely weak ionising effect**.
Penetration	Due to their strong ionising ability alpha particles can only penetrate a few centimetres in air. They can pass through very thin gold foil but cannot pass through thin card.	Beta particles are consequently **much more penetrating** than alpha particles and can travel up to 100 cm in air, pass through card and penetrate sheets of aluminium up to several millimetres thick.	Gamma radiation can **penetrate** several metres of concrete or several centimetres of lead, and is virtually unaffected by air.
Effect of a magnetic field (magnetic field into the page)			
Decay equation	$^A_ZX \longrightarrow\ ^{A-4}_{Z-2}Y\ +\ ^4_2\alpha\ +$ energy Parent nuclide Daughter nuclide Alpha particle	$^A_ZX \longrightarrow\ ^A_{Z+1}Y\ +\ ^0_{-1}\beta\ +$ energy Parent nuclide Daughter nuclide Beta particle	$^A_ZX^* \longrightarrow\ ^A_ZX\ +\ ^0_0\gamma$ Excited nuclide Unexcited nuclide Gamma radiation Often emitted at the same time as other forms of ionising radiation

Radioactive decay series

If the daughter nuclide produced by a radioactive decay is unstable it will also emit radiation to form a new daughter nuclide. The process of radioactive decay continues until a stable daughter nuclide is reached, and is referred to as a **radioactive decay series**.

Detecting radiation

The radiation emitted during a radioactive decay is invisible but it can be detected by sensors that react to the ionising properties of the radiation, e.g. photographic plates, Geiger-Müller (GM) tubes, cloud chambers.

Half-life

The emission of radiation is spontaneous and random so the 'life time' of a radioactive sample cannot be stated, as it is impossible to predict when the last nuclei will decay. If a graph of the number of nuclei is plotted against time, then the time taken for the sample to halve can be determined. The **half-life** ($t_{1/2}$) of a sample will always take the same time to halve regardless of how many atoms are in the sample. An example of a decay graph is shown below.

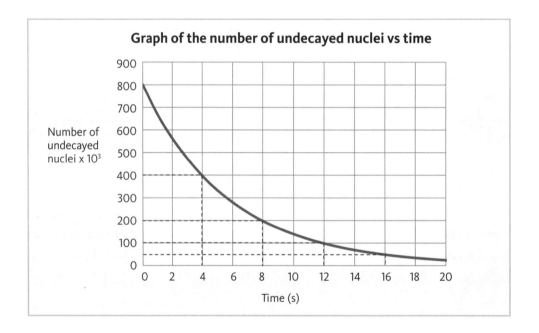

The half-life can also be determined mathematically as follows:

$$\text{number of nuclei remaining after } n \text{ half-lives} = \frac{\text{original number of nuclei}}{2^{(n \text{ half-lives})}}$$

Radioactive dating

Scientists can use their knowledge of the half-lives of different radionuclides to determine the age of a material, by comparing the amount of parent nuclei in a sample to the amount of its decay products. This technique is called **radioactive** or **radiometric dating**.

ISBN: 9780170195997

Uses and dangers of radiation

Smoke detectors, luminescent watch dials, nuclear medicine, sterilising, food preservation, machinery design and gene modification are all examples of the way we use radioactive materials in our homes and in medicine, industry and agriculture.

Radioactive materials also pose a health hazard if not used correctly. In humans, high doses of radiation will result in death due to the damage caused to the bone marrow, the digestive tract, the nervous system, the reproductive system and the eyes. Moderate doses are not immediately fatal. The body can replace damaged cells, however in the long term there is a significant increase in the risk of cancer.

Low doses have little effect on the human body, but doses can be cumulative and it is important that people are aware of their ongoing exposure, particularly people whose jobs place them in regular contact with radioactive materials, e.g. radiotherapists.

Background radiation

During our daily lives we are all being exposed to some ionising radiation in the form of cosmic rays from outer space, gamma radiation from radioactive rocks in the earth and building materials, the air we breathe and the food we eat. This **background radiation** is very low dose and there is currently no evidence to suggest that it adversely affects a person's health, however all radiation exposure should be kept to a minimum.

 ## Worked example: Fossilised tree

A preserved kauri tree stump is dug up and tested to determine how old it is. A 50 g sample of the tree is found to have an activity of 84 counts per minute, but a new 50 g sample of a kauri tree has an activity of 564 counts per minute.

Given that the half-life of carbon-14 is 5700 years, determine the age of the sample by graphing or otherwise.

Solution (by graphing)
By repeatedly dividing the count rate by two we find that the age lies somewhere between two and three half-lives.

Age (yrs)	Count rate
0	564
5700	282
11400	141
17100	71

Draw a graph with three half-lives on the x-axis and 600 counts per minute up the y-axis.

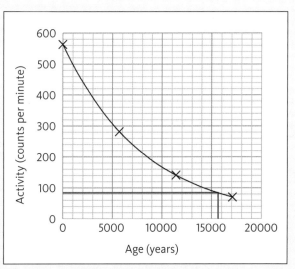

Draw interpolation lines from 84 on the y-axis to the graph line, then down to the x axis and read off the value. Age of sample ≈ 15 700 years old.

ISBN: 9780170195997

Exercise 5B

1   Lead-185 and lead-206 are both isotopes of lead, but lead-185 is an **unstable nuclide** and decays
to form mercury, as shown below.

$$^{185}_{82}Pb \longrightarrow \ ^{181}_{80}Hg \ + \ ^{A}_{Z}X$$

a   Complete the nuclear equation above by identifying the missing numbers and the radiation
that is emitted as a result of this reaction.

A =   Z =   X =

b   State the physics principles that you have used to write the nuclear equation.

_____

_____

c   Explain what is meant the term **unstable nuclide**. In your answer you should:
   •   describe the term nuclide
   •   describe the three conditions that could result in a nuclide being unstable
   •   explain why emission of radiation increases the stability of the nuclide.

_____

_____

_____

_____

_____

_____

_____

_____

_____

_____

_____

_____

_____

_____

_____

_____

_____

_____

_____

ISBN: 9780170195997

d Lead-185 has a half-life of **6 s**. If a sample of lead-185 has a mass of **1.20 kg**, use a graph (or by calculation) to determine the mass of lead remaining after 15 s.

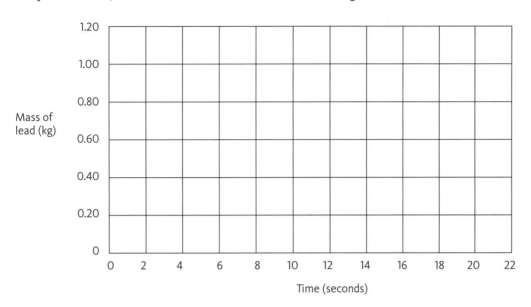

_____

_____

_____

_____

_____

_____

e The radiation emitted by lead as it decays can be detected using a Geiger counter. Describe and explain what will happen to the number of radiation emissions each second as the sample decays.

_____

_____

_____

_____

_____

_____

2 Rutherford demonstrated artificial radioactivity by firing alpha particles at nitrogen atoms. The nuclear equation below shows the reaction.

$$^{4}_{2}\alpha \; + \; ^{14}_{7}N \; \longrightarrow \; ^{17}_{8}O \; + \; ^{A}_{Z}X$$

a Complete the nuclear equation above by identifying the missing numbers and the nuclide that is emitted as a result of this reaction.

A =   Z =   X =

b How many protons and neutrons are found in the nucleus of an oxygen-17 atom?

_____

_____

c How many electrons will be found in an oxygen-17 atom? Explain your reasoning.

_____

_____

_____

_____

3 Cobalt-60 ($^{60}_{27}$Co) is an unstable nuclide of cobalt that can be created by the artificial nuclear reaction shown below.

$$^{59}_{27}\text{Co} + {}^{\square}_{\square}\square \longrightarrow {}^{60}_{27}\text{Co}$$

a Complete the nuclear equation above and identify the particle that has been added.

_____

_____

b State the physics principles that you have used to write the nuclear equation.

_____

_____

c Cobalt-60 has a half-life of approximately **5.25 years** and decays by emitting a beta particle and gamma radiation. Complete the nuclear equation below and use the periodic table in the appendix to identify the daughter nuclide that is formed.

$$^{60}_{27}\text{Co} \longrightarrow {}^{\square}_{\square}\square + {}^{0}_{-1}\beta + {}^{0}_{0}\gamma$$

d Explain why cobalt-60 is used in hospitals for sterilising medical equipment.

_____

_____

_____

_____

_____

e Explain what is meant by the term half-life, and why it is used to measure unstable nuclides.

_____

_____

_____

_____

ISBN: 9780170195997

f A sealed canister of mass 44 g contains 256 g of cobalt-60. Use a graph (or other method) to determine the time taken for the number of grams of cobalt-60 to decrease to only 25 g.

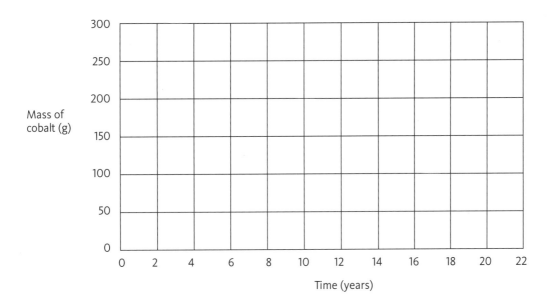

Mass of cobalt (g) vs Time (years)

_____

_____

_____

g Determine the mass of the canister and its contents after two half-lives. Explain your answer.

_____

_____

_____

_____

_____

h The beta particles that are emitted are often referred to as **nuclear electrons.** They are emitted at a range of different speeds. Discuss the term nuclear electron. In your answer you should:
- write a nuclear equation to show how a beta particle is formed
- consider how the decay obeys the laws of conservation
- explain how the emission increases the stability of the nuclide.

_____

_____

_____

_____

_____

_____

_____

ISBN: 9780170195997

4 An unknown parent nuclide decays to form radium-226, emitting two types of radiation in the process. To determine the type of radiations given off, different absorbers are placed between the unknown parent nuclide and the detector, and the results recorded in the table below.

a Use the table to determine the type of radiations given off. Explain your answers in terms of the properties of the two types of emission.

Absorber	Count rate (per minute)
No absorber	8460
Paper	320
Card	307
Thick aluminium	301
Thick lead	13

_____

_____

_____

_____

_____

_____

b Using your answers to the previous question, and with the aid of the periodic table in the appendix, complete the nuclear equation below to determine the parent nuclide that decayed to form radium-226 ($^{226}_{88}\text{Ra}$).

$$\boxed{}\boxed{} \longrightarrow \ ^{226}_{88}\text{Ra} \ + \ \boxed{}\boxed{} \ + \ \boxed{}\boxed{}$$

c Radium-226 is also unstable and decays to form radon-222. The graph below shows the change in mass of radium-226 against time. Use the graph to determine the half-life of radium-226.

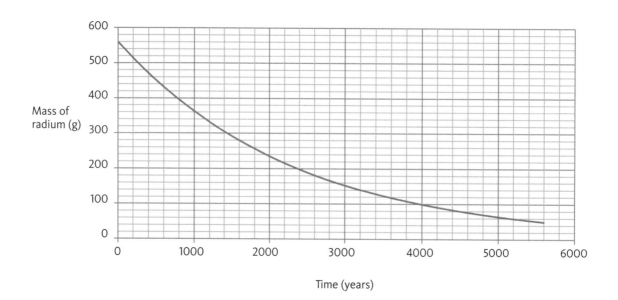

Mass of radium (g) vs Time (years)

ISBN: 9780170195997

_____

_____

_____

_____

_____

_____

_____

_____

_____

_____

5 Every day we are exposed to ionising radiation from our surroundings, or from various activities such as having an X-ray at the dentist. Natural radiation from the rocks under our homes accounts for 39% of the total background radiation to which we are exposed. Most of this radiation comes from radon-222, a radioactive gas given off as a decay product of uranium – it seeps into our houses where it decays, emitting an alpha particle.

a Complete the nuclear equation below by determining the missing numbers, and use the periodic table in Appendix 3 to identify the daughter nuclide that is formed.

$$^{222}_{86}\text{Ra} \longrightarrow \,^{A}_{Z}\text{X} + \,^{4}_{2}\alpha$$

A =	Z =	X =

In certain areas it is important to measure the amount of radon gas present in the air, especially inside poorly ventilated buildings. This is done by taking samples and sending them to a lab. The half-life of radon-222 is 3.8 days but it takes 19 days for the sample to reach the laboratory and be tested.

b Calculate the mass of radon-222 that would have been present in the sample when it was first taken, given that the tests found only 0.03 g present.

_____

_____

_____

_____

c Using your answer to the previous question, determine the mass of the daughter nuclide present in the sample jar when it was tested.

_____

_____

Up to 13% of our background radiation comes via gamma rays from space. Both alpha radiation and gamma radiation are described as forms of ionising radiation.

d   Discuss why alpha radiation emitted by the radon gas in the air we breathe poses a greater health risk than gamma radiation from space.

_____

_____

_____

_____

_____

_____

_____

6   Technetium-99 metastable is an isotope used in medical scanning. It is injected into a patient's body and emits gamma rays that can be detected by a special camera. It has a half-life of 6.0 hours, and is not retained in the human body for more than one day. Explain how these features make it ideal for use in medical imaging. In your answer you should:
- compare and contrast the ionising effects of alpha, beta and gamma rays
- the effect of radiation on living cells
- the importance of the half-life and how the body stores the isotope.

_____

_____

_____

_____

_____

_____

_____

_____

_____

_____

7   A patient is injected with 2.0 g sample of radioisotope iodine-131 to treat a serious thyroid illness. Iodine-131 has a half-life of 8.0 days and decays by emitting a beta particle.

a   Complete the nuclear equation below by determining the missing numbers, and use the periodic table in Appendix 3 to identify the daughter nuclide that is produced.

$$^{131}_{53}\text{I} \longrightarrow {}^{A}_{Z}\text{X} + {}^{0}_{-1}\beta$$

A = _____   Z = _____   X = _____

ISBN: 9780170195997

b The human body stores iodine for up to 136 days. Determine how many grams of iodine-131 will be left after 136 days.

_____

_____

_____

_____

Doctors, nurses and technicians who work with ionising radiation on a daily basis have to take precautions to avoid over-exposure. One way is to wear special dosimeter badges. The dosimeter is made from a photographic film covered by four windows containing copper, aluminium and plastic. Copper is three times denser than aluminium, and 20 times denser than the plastic. At the end of the day the film is developed, and any regions that have been exposed to radiation will have gone dark.

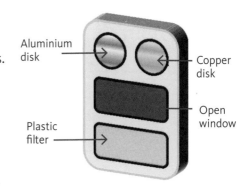

c Study the badges and identify whether the badges have been exposed to alpha, beta or gamma radiation.

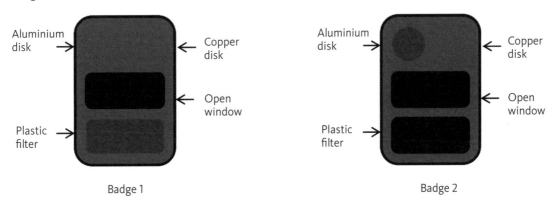

Badge 1   Badge 2

_____

_____

_____

d Explain how the dosimeter is able to measure the amount and type of radiation. In your answer you should consider:
 • the purpose of the open window on the badge
 • the reason for having different windows
 • the darkness of the film.

_____

_____

_____

_____

_____

_____

_____

_____

_____

_____

_____

_____

e   Suggest three alternative precautions that radiation workers should take to reduce their risk of over-exposure.

_____

_____

_____

_____

_____

_____

f   Describe the possible side effects of low level exposure to ionising radiation.

_____

_____

_____

_____

8   A preserved kauri tree stump is dug up and a 70 g sample tested using carbon dating to determine how old it is. A 210 g sample taken from a new tree is also tested. The results are shown in the table at right, along with the background radiation count.

Measurement	Average activity (counts per minute)
Background	8.0
Preserved tree sample (mass = 70 g)	9.0
New tree sample (mass = 210 g)	26.0

a   Describe the main sources of background radiation, and explain why it is important to measure the amount of background radiation before carrying out the experiment.

_____

_____

_____

_____

_____

_____

_____

_____

_____

ISBN: 9780170195997

b Given that the half-life of carbon-14 is 5700 years, determine the age of the sample by graphing (or otherwise).

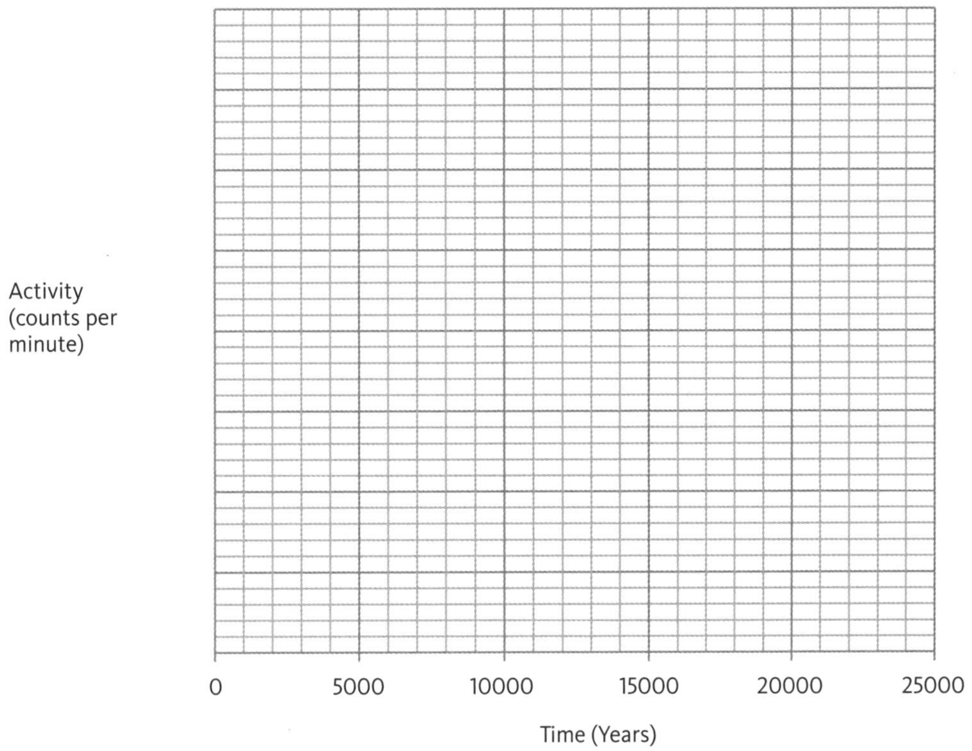

Activity
(counts per
minute)

0   5000   10000   15000   20000   25000

Time (Years)

_____

_____

_____

_____

_____

_____

_____

The amount of radiation emitted by the wood sample is measured using a Geiger counter, a device designed to detect ionising radiation.

c Explain what is meant by the term ionising radiation.

_____

_____

_____

_____

_____

_____

_____

_____

When a magnet is placed between the wood sample and the Geiger counter it causes the radiation to be deflected.

d State what this suggests about the emitted radiation, and explain how the equipment could be used to identify the type of radiation being emitted.

_____

_____

_____

_____

_____

_____

e Describe and explain an alternative test that could be carried out to determine the type of radiation being emitted.

_____

_____

_____

_____

_____

_____

9 Smoke detectors are used in homes and offices to alert people to the presence of smoke. Ionisation smoke detectors are inexpensive and very effective at detecting small amounts of smoke produced by flaming fires. They contain a small amount of americium-241, which has a half-life of 432 years, and is a good source of alpha particles.

a Complete the nuclear equation below by determining the missing numbers, and use the appendix to identify the daughter nuclide that is produced.

$$^{241}_{95}\text{Am} \longrightarrow ^{A}_{Z}\text{X} + ^{4}_{2}\alpha$$

A =   Z =   X =

b By considering the properties of the alpha, beta and gamma radiation, explain why alpha emitters are used to ionise the air between the charged plates inside the smoke alarm.

_____

_____

_____

_____

_____

_____

ISBN: 9780170195997

c   The smoke detector will work correctly provided the activity of the americium-241 is greater than 92% of the original activity when the alarm was first manufactured. By drawing a graph or otherwise, determine the effective working life of a smoke detector.

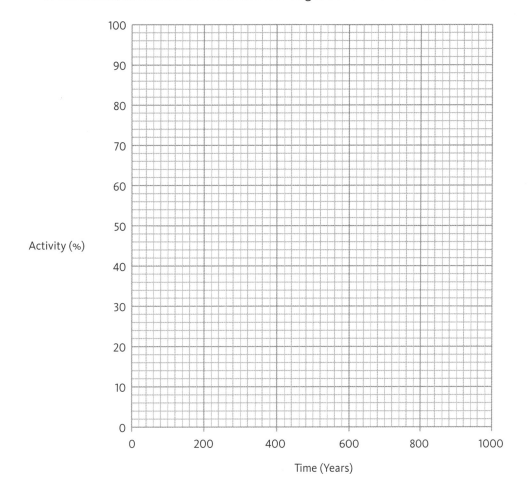

_____

_____

_____

_____

_____

_____

_____

_____

_____

_____

5.3   Radioactivity

Nuclear fission

The nucleus of an atom is held together by the **strong nuclear** force that acts between all the nucleons, and which is strong enough to overcome the electrostatic repulsion that is pushing the protons apart. The strong force has very short range, so small atoms tend to be very stable.

Nuclear fission is the process whereby the nucleus of an atom splits into smaller parts (called **fission fragments**). The process also releases high energy neutrons, gamma radiation and large amounts of energy.

Induced fission occurs when a heavy nuclide splits due to the addition of a neutron. This causes the nuclide to become unstable. The induced fission of uranium-235 is shown below.

$$^{235}_{92}U + ^{1}_{0}n \longrightarrow ^{236}_{92}U \longrightarrow ^{141}_{56}Ba + ^{92}_{36}Kr + 3^{1}_{0}n + energy$$

The nucleon and proton numbers balance in each part of the reaction, however, the mass at the start of the reaction is slightly greater than the mass at the end of the reaction. The difference in mass (Δm) is found as follows:

$$\Delta m = m_{final} - m_{initial}$$

Mass and energy

Albert Einstein proposed that mass and energy are related by the formula:

$$\begin{array}{ccccc} \textbf{energy} & = & \textbf{mass} & \times & \textbf{(speed of light)}^2 \\ \text{(J)} & & \text{(kg)} & & \text{(m s}^{-1})^2 \end{array}$$

Expressed mathematically:

$$E = mc^2$$

Where the speed of light:

$$c = 3.00 \times 10^8 \text{ m s}^{-1}$$

Einstein's theory means that instead of just considering the conservation of mass OR the conservation of energy, we must consider the conservation of mass-energy together.

ISBN: 9780170195997

Chain reactions

The fission of uranium-235 results in the emission of two or three fast moving high energy neutrons. Provided that there is more than a certain critical mass, these neutrons may result in further fissions in neighbouring uranium-235 nuclides, producing more energy and emitting even more neutrons – in other words, a chain reaction.

Nuclear power stations

The diagram below shows the various parts of a pressurised water reactor (PWR).

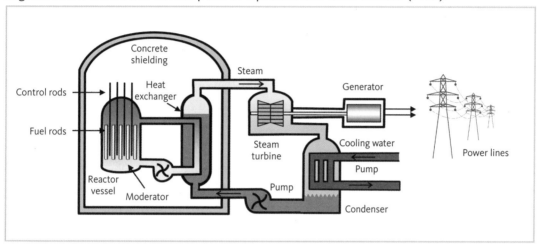

Part	Purpose
Fuel rods	Pellets of enriched uranium dioxide (97% is uranium-238, 3% is uranium-235) stored in long thin rods to increase the surface area.
Reactor vessel	Heavily shielded pressurised container that holds the fuel rods, control rods and moderator.
Moderator	Surrounds the fuel rods and absorbs some of the kinetic energy of the fast moving neutrons to reduce their speed to that of gas particles at normal temperatures (such neutrons are therefore referred to as **thermal neutrons**), e.g. water, heavy water, graphite.
Control rods	The control rods absorb neutrons so the number of neutrons available to cause fission in the uranium-235 decreases, reducing the rate of the fission reactions, e.g. boron, cadmium.
Heat exchange	Transfer the heat energy from the moderator to normal water to create steam to turn the turbines.

ISBN: 9780170195997

The power output (see Work, energy and power on pages 141–142) of the reactor can be determined using the power formula:

$$\text{power} = \frac{\text{energy changed (J)}}{\text{time taken (s)}}$$
$$(\text{J s}^{-1})$$

Expressed mathematically:

$$P = \frac{\Delta E}{t}$$

Radioactive waste

Some of the waste from reactors can be reprocessed or used for other applications, but most of the waste has to be stored for hundreds of years due to the long half-lives of the fission fragments produced during the fission process. The waste is first stored in large cooling ponds at the reactor site, then after about 10 to 20 years the remaining radioactive waste is sealed in concrete and steel drums, and buried deep underground or dropped to the bottom of the sea.

Nuclear fusion

Fusion occurs when light nuclides collide and come close enough for the strong force to overcome the electrostatic repulsion (due to the positive charge), and fuse (stick) the two nuclei together to form a larger nucleus. One possible **fusion reaction** is shown below, involving two isotopes of hydrogen – hydrogen-2 (deuterium) and hydrogen-3 (tritium) – to form helium:

$$^{2}_{1}\text{H} + {}^{3}_{1}\text{H} \longrightarrow {}^{4}_{2}\text{He} + {}^{1}_{0}\text{n}$$

Fusion reactions may provide us with an alternative source of energy for the future, but there are many significant problems involved in trying to create and maintain the extreme conditions necessary for sustainable nuclear fusion.

As a power source, fusion has a number of advantages over fission:
- Much greater energy yield per kilogram.
- Readily available fuel supply.
- No toxic or highly radioactive waste materials produced.

As with all types of power station there will be some waste produced, especially waste heat energy, which cannot be turned into electricity and ends up being released into the environment.

Worked example: Energy from fusion

Calculate the energy released when deuterium and tritium undergo fusion using the formula and data in the table shown below. (*Note:* $c = 3.00 \times 10^8 \text{ m s}^{-1}$)

$$^{2}_{1}\text{H} + {}^{3}_{1}\text{H} \longrightarrow {}^{4}_{2}\text{He} + {}^{1}_{0}\text{n}$$

Nuclides	Mass (kg)
$^{2}_{1}\text{H}$	3.34358×10^{-27}
$^{3}_{1}\text{H}$	5.00736×10^{-27}
$^{4}_{2}\text{He}$	6.64466×10^{-27}
$^{1}_{0}\text{n}$	1.67493×10^{-27}

ISBN: 9780170195997

Solution

Given	$m_{initial}$	= $3.34358 \times 10^{-27} + 5.00736 \times 10^{-27} = 8.35094 \times 10^{-27}$ kg
	m_{final}	= $6.64466 \times 10^{-27} + 1.67493 \times 10^{-27} = 8.31959 \times 10^{-27}$ kg

Unknown $E = ?$

Equations $\Delta m = m_{final} - m_{initial}$
$E = mc^2$

Substitute $\Delta m = 8.31959 \times 10^{-27m} - 8.35094 \times 10^{-27}$
$\Delta m = -0.03135 \times 10^{-27}$
$E = 0.03135 \times 10^{-27} \times (3.00 \times 10^8)^2$

Solve $E = 2.8215 \times 10^{-12}$ J
$E = 2.82 \times 10^{-12}$ J (3 s.f.)

Exercise 5C

Take the speed of light, c = 3.00 × 10⁸ m s⁻¹ when required.

1   The diagram below shows an induced fission reaction that could take place inside a nuclear reactor.

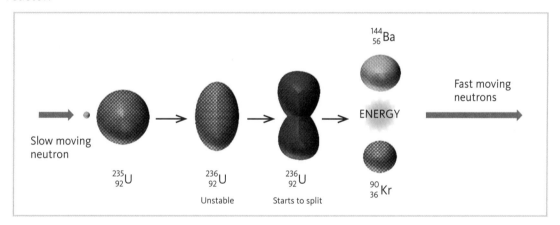

a   Use the information in the diagram to write a balanced nuclear equation in the box below, and determine the number of neutrons released by the reaction.

When the slow moving neutron strikes the spherical uranium-235 nucleus it joins with it, producing uranium-236, which becomes elongated like a rugby ball before separating into two parts and releasing energy.

b By considering the forces acting on the nucleus, explain why the spherical nucleus is stable but the elongated nucleus is unstable and breaks apart. In your answer you should consider:
- the **strength** of the strong force compared to the electrostatic force
- the **range** of the strong force compared to the electrostatic force.

_____

_____

_____

_____

_____

_____

_____

c Explain the conditions necessary for a sustainable chain reaction to take place in the reactor. In your answer you should consider:
- how the high speed neutrons released by the reaction above can become slow moving neutrons inside the reactor
- the amount of uranium-235 present in the reactor
- the role of the moderator and control rods.

_____

_____

_____

_____

_____

_____

_____

_____

_____

_____

_____

_____

_____

_____

_____

_____

ISBN: 9780170195997

d Use the table at right and your balanced nuclear equation from **a**, to calculate the mass of the reactants and the products, and show that a single uranium nucleus undergoing fission releases 28.8×10^{-12} J of energy.

Nuclides	Mass (kg)
$^{235}_{92}U$	390.216×10^{-27}
$^{1}_{0}n$	1.67493×10^{-27}
$^{144}_{56}Ba$	238.939×10^{-27}
$^{90}_{36}Kr$	149.282×10^{-27}

_____

_____

_____

_____

_____

_____

_____

_____

_____

e Calculate the mass of uranium-235 required to produce 2.000 GJ of energy. Give your answer in kilograms.

_____

_____

_____

_____

2 A pressurised water reactor (PWR) is filled with 110 tonnes of enriched uranium dioxide (1 tonne = 1000 kg). Each 1 kg of enriched uranium dioxide contains 2.22×10^{24} atoms of uranium, of which 96% are uranium-238 and 4.00% are uranium-235 nuclides. The initial step in the nuclear fission reaction involves a **thermal neutron** colliding with a uranium-235 nuclide and forming uranium-236.

a Explain what is meant by the term thermal neutron.

_____

_____

_____

_____

b Determine the number of uranium nuclei inside the reactor, and hence calculate the number of uranium-235 nuclei.

_____

_____

_____

_____

c Using the table at right, calculate the mass of the reactants and the products and show that 27.800×10^{-12} J of energy are released by a single uranium nuclei undergoing fission.

Nuclides	Mass (kg)
$^{235}_{92}U$	390.216×10^{-27}
$^{1}_{0}n$	1.67493×10^{-27}
$^{144}_{56}Ba$	238.939×10^{-27}
$^{89}_{36}Kr$	147.618×10^{-27}

$$^{235}_{92}U + {}^{1}_{0}n \longrightarrow {}^{144}_{56}Ba + {}^{89}_{36}Kr + 3{}^{1}_{0}n + \text{energy}$$

_____

_____

_____

_____

_____

_____

_____

_____

_____

The power station is required to produce an average of 54.3 GW of electrical energy.

d Calculate how long the reactor could potentially operate at an average power output of 50 GW. State any assumptions you have made to reach your answer.

_____

_____

_____

_____

_____

_____

_____

3 Fast breeder reactors (FBR) use uranium-235 as the main fuel but they are able to make their own fuel by converting the uranium-238 into plutonium-239, which can also undergo fission to release energy. When a uranium-238 nuclide captures a fast moving neutron it forms uranium-239, which then decays by emitting radiation to form an isotope of plutonium-239.

a Complete the nuclear equation to determine the radiation emitted during the nuclear reaction.

$$^{238}_{92}U + {}^{1}_{0}n \longrightarrow {}^{239}_{92}U$$

Then:

$$^{239}_{92}U \longrightarrow {}^{239}_{94}Pu + \boxed{} + \text{energy}$$

ISBN: 9780170195997

The plutonium will then undergo fission when it captures a neutron. A possible nuclear reaction is shown below.

$$^{239}_{94}\text{Pu} + {}^{1}_{0}\text{n} \longrightarrow {}^{120}_{47}\text{Ag} + {}^{117}_{47}\text{Ag} + 3{}^{1}_{0}\text{n} + \text{energy}$$

b With reference to the reactions above define the terms element, nuclide and isotope.

_____

_____

_____

_____

_____

_____

_____

c Use the table at right to calculate the mass of the reactants and the products, and so determine the energy released by a single plutonium nuclide undergoing fission.

Nuclides	Mass (kg)
$^{239}_{94}\text{Pu}$	396.870×10^{-27}
$^{1}_{0}\text{n}$	1.67493×10^{-27}
$^{120}_{47}\text{Ag}$	199.087×10^{-27}
$^{117}_{47}\text{Ag}$	194.094×10^{-27}

_____

_____

_____

_____

_____

_____

_____

_____

_____

d State the physics principle you used to solve the above problem.

_____

_____

e Plutonium has a much lower critical mass than uranium, so less fuel is required in a reactor that uses plutonium as a fuel. Explain what is meant by the term 'critical mass'.

_____

_____

_____

_____

4 Hydrogen is the lightest element and has many isotopes. The three most stable nuclides are hydrogen-1 (known as protium), hydrogen-2 (known as deuterium) and hydrogen-3 (known as tritium).

a Explain the meaning of the terms 'isotope' and 'nuclide'.

_____

_____

_____

_____

_____

_____

_____

_____

The diagram below shows a nuclear reaction that could take place inside a fusion reactor. When the deuterium and tritium collide with sufficient speed they fuse together to produce helium, emit a neutron and release energy.

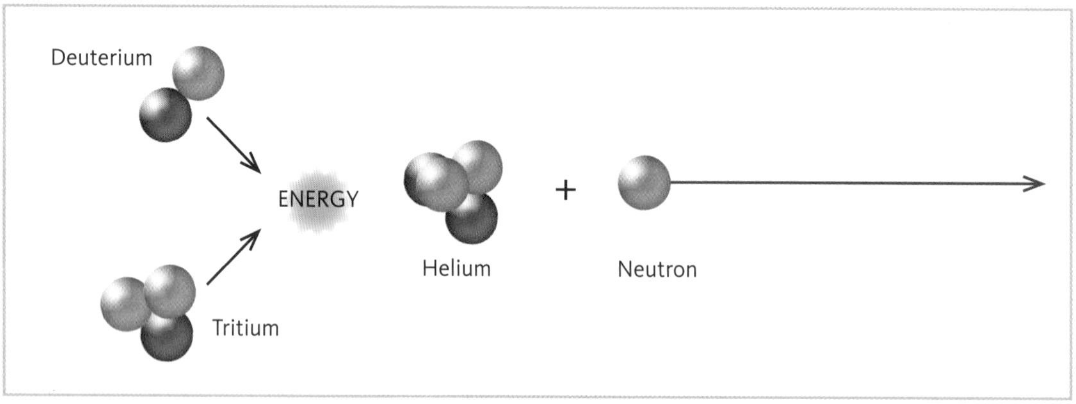

b Use the information in the diagram to write a balanced nuclear equation in the box below.

c By considering the forces acting between the nuclei, explain why a large amount of energy is required to cause the nuclei to fuse. In your answer you should consider:
  - the **strength** of the strong force compared to the electrostatic force
  - the **range** of the strong force compared to the electrostatic force
  - the relationship between the force and the energy supplied to start the reaction.

_____

_____

_____

ISBN: 9780170195997

d Using the table at right, calculate the mass of the reactants and the products, and show that a single fusion reaction will release 2.82×10^{-12} J of energy.

Nuclides	Mass (kg)
$_{1}^{2}\text{H}$	3.34358×10^{-27}
$_{1}^{3}\text{H}$	5.00736×10^{-27}
$_{2}^{4}\text{He}$	6.64466×10^{-27}
$_{0}^{1}\text{n}$	1.67493×10^{-27}

e Calculate the total number of fusion reactions required to produce 2.000 GJ of energy, hence determine the total mass of the reactants. Give your answer in kilograms.

The Sun produces energy by nuclear fusion and it is radiated away at a rate of 3.8×10^{26} W.

f Compare your answer to **4e** with your answer to **1e**, and comment on the difference in fuel consumption in fisson and fusion reactions.

ISBN: 9780170195997

g   Calculate the mass of deuterium and tritium that would have to be reacted each day to sustain this energy output.

_____

_____

_____

_____

_____

_____

_____

_____

_____

_____

ISBN: 9780170195997

6

Electricity

Achievement Standard AS91173 (P2.6) requires students to demonstrate their knowledge and understanding of phenomena, concepts or principles related to electricity and electromagnetism. To demonstrate understanding of electricity and electromagnetism is worth 6 credits and is assessed externally.

Static electricity
- Uniform electric fields
- Electric field strength
- Force on a charge in an electric field
- Electric potential energy
- Work done on a charge moving in an electric field.

DC electricity
- Parallel circuits with resistive component(s) in series with the source
- Circuit diagrams
- Voltage
- Current
- Resistance
- Energy
- Power.

Electromagnetism
- Force on a current carrying conductor in a magnetic field
- Force on charged particles moving in a magnetic field
- Induced voltage generated across a straight conductor moving in a uniform magnetic field.

The following relationships are provided in the assessment.

$$E = \frac{V}{d} \qquad F = Eq \qquad \Delta E_p = qEd \qquad E_k = \tfrac{1}{2}mv^2 \qquad F = BIL \qquad F = Bqv$$

$$V = BvL \qquad I = \frac{q}{t} \qquad V = \frac{\Delta E}{q} \qquad V = IR \qquad P = IV \qquad P = \frac{\Delta E}{t}$$

$$R_T = R_1 + R_2 + \dots \qquad \frac{1}{R_T} = \frac{1}{R_1} + \frac{1}{R_2} + \dots$$

ISBN: 9780170195997

6.1 Electrostatics

Charge (Q)

Charge is a fundamental property of matter. It occurs in two forms that are described as positive and negative. The **quantity of charge** (Q) on an object is measured in **coulombs** (C). The charge on electrons and protons is extremely small:

> charge on an electron $q_e = -1.60 \times 10^{-19}$ C
>
> charge on a proton $q_p = +1.60 \times 10^{-19}$ C

Electric fields

An electric field is a region in space in which an electric charge will experience a force of attraction or repulsion. By convention the direction of an electric field at any point is defined as the direction a force would cause a small positive charge to move.

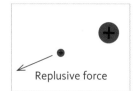

Replusive force

Electric fields can be represented by drawing **lines of force**, with arrows on them to show the direction of the electric field. These lines of force are also known as **electric field lines** and always leave or enter a surface at right angles. The number of electric field lines in an area indicates how strong the electric field is at that point.

Uniform electric fields

The electric field that forms between two oppositely charged metal plates is illustrated at right. The field lines are **parallel and equally spaced** except near the edges of the plates. This means that the strength of the electric field will be the same at every point in the middle (shaded blue) and can therefore be described as a **uniform field**.

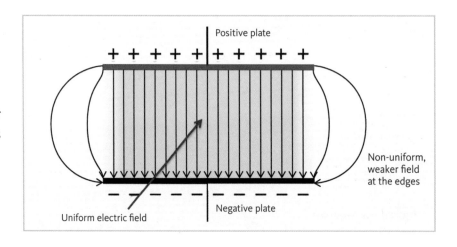

Positive plate

Non-uniform, weaker field at the edges

Uniform electric field

Negative plate

Electric field strength (E)

The strength of the electric field that forms between the two plates depends upon the difference in the voltage between the two plates (known as **potential difference** or V, see page 198), and how far apart the plates are from each other.

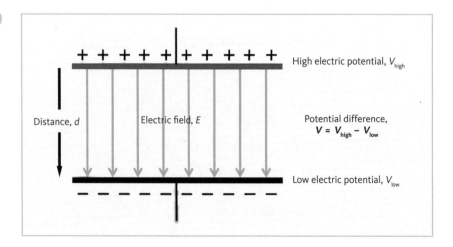

High electric potential, V_{high}

Distance, d

Electric field, E

Potential difference, $V = V_{high} - V_{low}$

Low electric potential, V_{low}

ISBN: 9780170195997

If there is a **potential difference** between two plates a **distance** (d) apart then the electric field **strength** (E) between the plates is found by:

$$\frac{\textbf{electric field strength}}{\textbf{(V m}^{-1}\textbf{)}} = \frac{\textbf{potential difference between the plates} \quad \text{(V)}}{\textbf{distance between the plates} \quad \text{(m)}}$$

Expressed mathematically:

$$E = \frac{V}{d}$$

Worked example: Electrical field strength

Two metal plates are placed opposite each other at a distance of 0.5 cm and have a potential difference of 1.50 kV. Calculate the strength of the electric field.

Solution

Given $V = 1.50 \times 10^3$ V
$d = 0.5$ cm $= 0.5 \times 10^{-2}$ m

Unknown $E = ?$

Equations $E = \dfrac{V}{d}$

Substitute $E = \dfrac{1.50 \times 10^3}{0.5 \times 10^{-2}}$

Solve $E = 3.0 \times 10^5$ V m^{-1} (1 s.f.)

Force (F) on a charge in an electric field

If a small positive **charge** (q) is placed in the uniform **electric field of strength** (E) it will experience a **force** (F), causing it to accelerate away from the positive plate and towards the negative plate. The size of the force on a charge can be calculated by:

$$\textbf{electric force} = \textbf{charge} \times \textbf{electric field strength}$$
$$\text{(N)} \qquad \text{(C)} \qquad \text{(N C}^{-1}\text{)}$$

Expressed mathematically:

$$F = qE$$

Worked example: Electrical force on a charged particle

A hydrogen ion of charge $+1.6 \times 10^{-19}$ C is moving in an electric field of strength 250 kN C^{-1}. Calculate the strength of the electric force acting on the ion.

Solution

Given	$q = +1.6 \times 10^{-19}$ C
	$E = 250 \times 10^{3}$ N C^{-1}
Unknown	$F = ?$
Equations	$F = qE$
Substitute	$F = (1.6 \times 10^{-19}) \times (250 \times 10^{3})$
Solve	$F = 4.0 \times 10^{-14}$ N (2 s.f.)

Work done and electrical potential energy

When a small **charge** (q), is placed in an electric field it will experience a **force** (F) of either attraction or repulsion, which will cause the charge to move a **distance** (d.) Whenever a force moves a distance, energy is changed and work is done. **Energy changed** (ΔE) is given by:

$$\Delta E = qEd$$

This states that:

$$\text{energy changed} = \text{charge} \times \text{electric field strength} \times \text{distance moved}$$
$$\text{(J)} \qquad\qquad \text{(C)} \qquad\qquad \text{(N C}^{-1}) \qquad\qquad \text{(m)}$$

The above represents the electrical potential energy changed as the charged particle moves, and can be otherwise expressed as:

$$\Delta E_{p} = qV$$

Which states further that:

$$\text{electric potential energy changed} = \text{charge} \times \text{potential difference}$$
$$\text{(J)} \qquad\qquad\qquad \text{(C)} \qquad\qquad \text{(V)}$$

Principle of conservation of energy

When an electric field does work on a charged particle it causes the particle to lose electrical potential energy and gain kinetic energy.

Electric fields can be used to accelerate particles up to very high speeds, which can be determined when using the equation:

$$E_{K} = \frac{1}{2}mv^{2}$$

ISBN: 9780170195997

Worked example: Velocity of a charged particle moving in an electric field

A helium ion of mass $m = 6.645 \times 10^{-27}$ kg and charge of $q = 3.2 \times 10^{-19}$ C is at rest when it experiences a force due to an electric field of strength 5.00×10^4 N C^{-1} causing it move 0.70 cm. By considering the energy gained by the particle, determine its final velocity.

Solution

Given

$m = 6.645 \times 10^{-27}$ kg
$q = 3.2 \times 10^{-19}$ C
$E = 5.00 \times 10^4$ NC^{-1}
$d = 0.7$ cm $= 0.7 \times 10^{-2}$ m

Unknown

$v = ?$

Equations

$\Delta E_p = qEd$ and $\Delta E_K = \dfrac{1}{2}mv^2$

Substitute

$\Delta E_p = 3.2 \times 10^{-19} \times 5.00 \times 10^4 \times 0.70 \times 10^{-2} = 1.12 \times 10^{-16}$ J
Assuming no energy is supplied or removed from the system we can apply the principle of conservation of energy, i.e. loss in E_p = gain in E_K
$\Delta E_K = 1.12 \times 10^{-16}$ J

$\Delta E_K = \dfrac{1}{2}mv^2$

$1.12 \times 10^{-16} = \dfrac{1}{2} \times 6.645 \times 10^{-27} \times v^2$

Solve

$v_f = \sqrt{\dfrac{1.12 \times 10^{-16}}{3.3225 \times 10^{-27}}}$

$v_f = \pm 1.8 \times 10^5$ m s^{-1} (2 s.f.)

Path of a moving charge in a field

When a positively charged particle is fired into an electric field, it experiences a **constant force** of repulsion from the positive plate causing it to follow a **parabolic path**.

When a negatively charged particle is fired into an electric field, it experiences a constant force of attraction towards the parabolic path in the **opposite direction**.

Once the charge leaves the field it travels in a straight line

Velocity, v

ISBN: 9780170195997

Exercise 6A

Charged particles, force equations and electric fields

Take the charge on an electron as q_e = -1.60 x 10^{-19} C.

1   Calculate the charge on a droplet of oil coated with 2.0×10^5 electrons.

_____

2   Calculate the number of electrons on a dust particle with a charge of -4.8 $\times$ 10^{-10} C.

_____

3   Draw **an arrow** to show the relative size and direction of the electric forces acting on the free
charges below and explain your answer.

a   Two free charged particles close to a strong fixed charge.

Free charge

Free charge

Fixed charge

_____

_____

_____

_____

_____

_____

b   Two free charged particles in between charged plates.

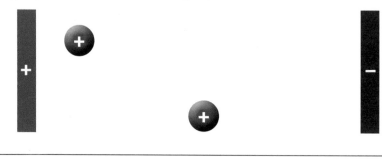

_____

_____

_____

4   Explain what is meant by the term **electric field**. Draw the shape of the electric field between
the plates in the diagram below.

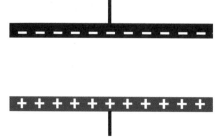

ISBN: 9780170195997

_____

_____

_____

5 Use the electric field strength equation $E = \dfrac{V}{d}$ to solve the following problems.

a Two plates are 0.06 m apart and have a potential difference of 1.34×10^3 V. Calculate the strength of the electric field.

_____

b An electric field of strength 180 kV m^{-1} is formed between two metal plates 3.0 cm apart. Calculate the potential difference between the plates.

_____

c Two metal plates have a potential difference of 120 kV across them, resulting in an electric field of strength 8.00×10^5 V m^{-1}. Calculate the distance between the plates.

_____

6 Use the electric force equation $F = Eq$ to solve the following problems.

a A helium ion with a charge of 3.2×10^{-19} C is placed in an electric field of strength 80 kV m^{-1}. Calculate the size of the electric force.

_____

b An electron with a charge of -1.6×10^{-19} C is placed in an electric field of strength 50 kV m^{-1}.

i Calculate the size of the electric force.

_____

ii What does the negative sign tell you about the direction of the force?

_____

c A force of -3.6 µN acts on an oil drop of charge -1.08×10^{-10} C.

i Calculate the strength of the electric field.

_____

ii State two units for the electric field strength.

_____

d An ion experiences a force of 1.28×10^{-13} N while in an electric field of strength 400 kV m^{-1}. Calculate the charge on the ion.

_____

7 Use the electrical potential energy equation $\Delta E = qEd$ to solve the following problems.

a A hydrogen ion with a charge of 1.6×10^{-19} C is placed in an electric field of strength 40 kV m^{-1} causing it to move 3.0 cm. Calculate the change in electrical potential energy.

_____

_____

b A tiny droplet of water with a charge of -5.6 × 10⁻⁹ C is at rest between two metal plates when an electric field is briefly switched on, causing the droplet to move 1.2 cm and change 4.70 µJ of energy.

    i In which direction will the droplet move relative to the electric field direction?

_____

    ii Calculate the strength of the electric field.

_____

_____

c An electron with a charge of -1.6 × 10⁻¹⁹ C and a mass of 9.11 × 10⁻³¹ kg is accelerated from rest up to a speed 1.5 × 10⁷ m s⁻¹, by an electric field of strength 200 kV m⁻¹.

    i Calculate the gain in kinetic energy.

_____

_____

    ii With reference to the principle of conservation of energy, explain where the energy came from.

_____

_____

_____

_____

_____

    iii Determine how far the electron moves within the field.

_____

_____

_____

_____

8 A Van de Graaff generator produces a potential difference (voltage) of 3.00 × 10⁵ V between the dome and the earth. A metal plate is connected to the dome of the generator and placed 10.0 cm away from a second plate connected to the earth, as shown in the diagram below.

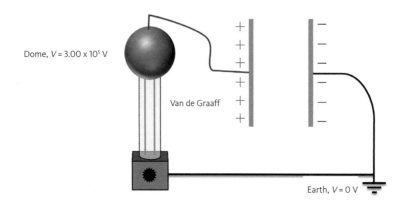

Dome, V = 3.00 x 10⁵ V

Van de Graaff

Earth, V = 0 V

ISBN: 9780170195997

a Draw the shape of the field on the previous diagram and explain its shape.

_____

_____

_____

_____

_____

_____

b Calculate the strength of the electric field between the plates and state the unit of electric field strength.

_____

_____

A lightweight aluminium ball with a mass of 15.0×10^{-2} kg is suspended from a string so that it hangs freely between the plates.

c On this diagram show the charge distribution on the metal ball when it is hanging in the middle of the two plates, and explain why the ball doesn't move.

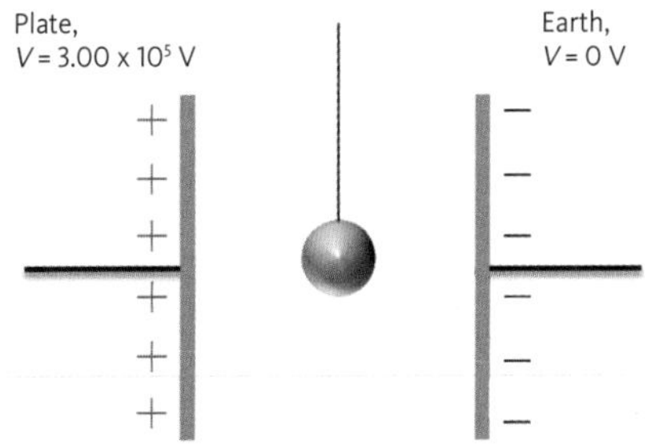

_____

_____

_____

_____

_____

_____

Plate,
$V = 3.00 \times 10^5$ V

$+$
$+$
$+$
$+$
$+$
$+$

Earth,
$V = 0$ V

$-$
$-$
$-$
$-$
$-$
$-$

The ball is moved to one side so it touches the positive plate and gains a charge of 1.0×10^{-6} C.

d Calculate the number of electrons that must have been transferred to produce this charge on the ball.

_____

_____

e Calculate the electric potential energy of the ball while it is in contact with the positive plate.

_____

_____

_____

ISBN: 9780170195997

f Explain how the answer to **e** can be used to calculate the kinetic energy of the ball when it reaches the opposite plate.

_____

_____

_____

g Determine the velocity at which the ball strikes the opposite plate.

_____

_____

_____

_____

_____

9 Cathode Ray Oscilloscopes (CRO) are often used to make measurements of electrical signals that are rapidly changing, for example, the alternating current flowing through a loud speaker. The image on the screen of the CRO is produced by an electron beam striking a fluorescent screen, as shown below. (The charge on an electron is = -1.60 × 10^{-19} C.)

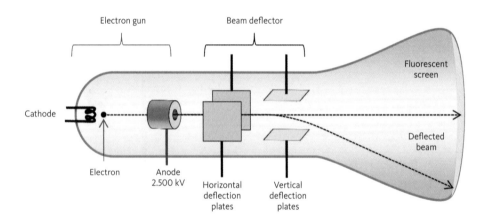

Part 1: The electron gun

At the end of the CRO is an electron gun. The electrons are emitted by the cathode and are accelerated from rest by the anode, which is maintained at a potential difference of 2.500 kV. When the electrons reach the anode they are moving at a speed of 2.973 × 10^7 m s^{-1}.

a On the diagram above draw an arrow to show the **direction** of electric field inside the electron gun (between the cathode and the anodes.)

ISBN: 9780170195997

b Show that 4.00×10^{-16} J of electrical potential energy are changed when an electron is accelerated from the cathode towards the anode.

_____

_____

_____

c Calculate the mass of an electron, clearly stating any laws of physics that you use to solve the problem.

_____

_____

_____

_____

_____

_____

Part 2: The beam deflector

The direction of the beam can be changed using the horizontal or vertical deflection plates. An electric field of strength 12.5 kV m⁻¹ is formed between the two vertical plates, which are 4.0 cm apart causing the electron beam to be deflected downwards, as shown in the diagram on page 192.

d Calculate the potential difference between the plates.

_____

_____

e Calculate the size of the electric force that acts on each electron as it passes between the plates.

_____

_____

f Describe and explain in detail the motion of the electrons while they travel between the plates, then on towards the screen.

_____

_____

_____

_____

_____

_____

_____

_____

10 To improve the quality and effectiveness of a paint spray gun, the nozzle at the end of it is attached to the negative terminal of a battery, so that as the paint leaves the gun it becomes negatively charged. The surface of the object being sprayed is connected to the positive terminal of the battery so that it attracts the paint droplets, resulting in more paint striking the surface and a more even spread of paint. This technique is very effective

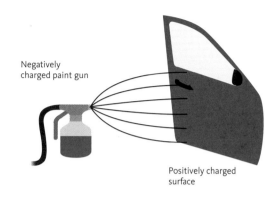

Negatively charged paint gun

Positively charged surface

when spraying cars. The diagram above shows the electric field lines which the paint droplets follow as they travel from the nozzle to a car door.

a Draw an arrow on the diagram to show the direction of the electric field between the paint gun's nozzle and the car door.

b Explain why this cannot be described as a uniform electric field.

_____

_____

_____

_____

_____

_____

_____

_____

Alice wants to study the forces acting on the charged paint droplets in a uniform electric field so she sets up two metal plates in the laboratory and injects paint droplets into the electric field between the plates. Her experimental set-up is shown below.

View from above

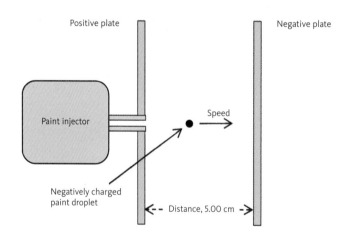

Positive plate

Negative plate

Paint injector

Speed

Negatively charged paint droplet

Distance, 5.00 cm

ISBN: 9780170195997

Each paint droplet has a mass of 8.9×10^{-10} kg and is injected into the electrical field at a speed of 15.0 m s^{-1}. Alice increases the potential difference between the plates until the paint droplets come to a stop just before they reach the negative plate. The droplets then accelerate back towards the positive plate. The paint droplets are coated with 5.21×10^6 electrons each with a charge of -1.6×10^{-19} C.

c   Draw an arrow on the diagram to show the direction of the electric **force** on the paint droplet, and describe the motion of the paint droplet as a result of this force.

_____

_____

d   Explain in terms of force and energy why the paint droplet comes to a stop just before it reaches the negative plate.

_____

_____

_____

_____

_____

_____

e   Calculate the size of the potential difference between the plates.

_____

_____

_____

_____

_____

_____

_____

11   Ingrid is investigating the relationship between electric field strength and the distance between two charged metal plates. She connects two plates to a power supply and then increases the distance between the plates. With each new distance she measures the strength of the electric field. The measurements are shown below.

a   Identify any anomalous values and calculate the averages for the data.

Distance (m) (±0.001 m division error)	Electric field strength (kVm⁻¹) (±0.005 kVm⁻¹ division error)			
	Trial 1	Trial 2	Trial 3	Average
0.0100	12.002	11.999	12.003	
0.0200	6.000	5.998	5.996	
0.0300	4.402	4.004	3.998	
0.0400	3.001	3.999	3.001	
0.0500	2.401	2.399	2.400	
0.0600	2.001	1.999	2.000	
0.0700	1.714	1.713	1.715	

b Draw a graph of electric field strength (y) against distance (x) using the data above, and draw on the line of best fit.

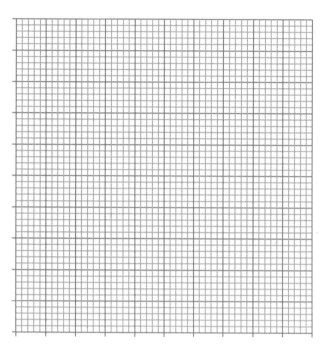

c State the relationship based upon the shape of the graph.

_____

_____

d Complete the last column of the table by processing the data. Include an appropriate quantity and unit.

e Plot a second graph using your processed data on the graph below.

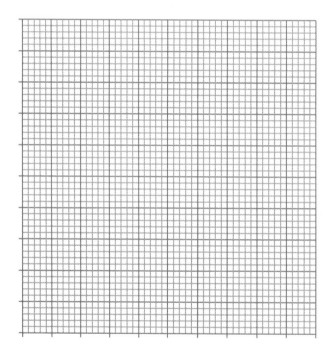

f Determine the gradient of the graph and the intercept and state the equation of the line. Provide units with all your values.

_____

_____

_____

g Compare the equation of the line to the theoretical equation $E = \dfrac{V}{d}$ and hence determine the supply voltage.

_____

_____

_____

ISBN: 9780170195997

6.2 Electric quantities

Conventional current (I)

Current is the flow of positive charges around a circuit each second. If a quantity of charge (Q) passes a given point in time (t), then a steady current (I) flows, hence:

$$\frac{\textbf{current}}{\text{(A)}} = \frac{\textbf{charge} \quad \text{(C)}}{\textbf{time} \quad \text{(s)}}$$

Expressed mathematically:

$$I = \frac{Q}{t}$$

Current is measured in **ampere** (A) (usually shortened to 'amps') using an ammeter or $-\!(\!A\!)\!-$. Ammeters are always connected in series (or in line) with a component, to measure the rate at which the charges enter or leave the component.

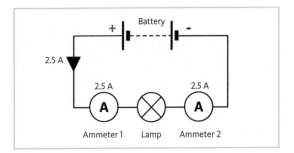

Conservation of charge

As charge is conserved, the rate at which charges enter and leave a component will be the same, so an ammeter will provide the same reading on either side of a component.

 Worked example: Current through a resistor

A charge of 30 C flows through a lamp every minute. Determine the size of the current.

Solution

Given $Q = 30$ C
        $t = 1$ min $= 60$ s

Unknown $I = ?$

Equations $I = \dfrac{Q}{t}$

Substitute $I = \dfrac{30}{60}$

Solve $I = 0.5$ A (2 s.f.)

Electrical potential difference or voltage (V)

Potential difference (V) is the amount of **electrical energy changed** (ΔE) to other forms of energy by **one coulomb of charge** (Q) moving from a point of high electrical potential energy to a point of lower electrical potential energy.

Potential difference or voltage can be expressed as an equation:

$$\frac{\textbf{voltage}}{\textbf{(V)}} = \frac{\textbf{energy changed}}{\textbf{charge}} \quad \frac{\text{(J)}}{\text{(C)}}$$

Expressed mathematically:

$$V = \frac{\Delta E}{Q}$$

Potential difference is measured in **volts** (V), using a voltmeter, or ⎯Ⓥ⎯. As voltage is 'used up' inside a resistor or transducer, voltmeters must be connected across (in parallel with) them to take 'before and after' readings.

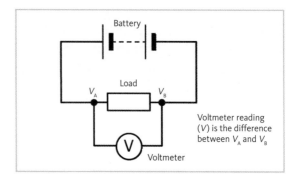

Voltmeter reading (V) is the difference between V_A and V_B

Worked example: Voltage across a battery

120.0 J of energy is converted from chemical to electrical energy by a battery when 18.0 C of charge flows through it. Calculate the potential difference across the battery.

Solution

Given

$Q = 18.0$ C
$\Delta E = 120.0$ J

Unknown

$V = ?$

Equations

$$V = \frac{\Delta E}{Q}$$

Substitute

$$V = \frac{120.0}{18.0}$$

Solve

$V = 6.67$ V (3 s.f.)

Ohm's Law and resistance (R)

Resistance is a measure of a component's opposition to the flow of electrical charge. Ohm's Law states that at constant temperature the voltage across a component is directly proportional to the current flowing through it.

Materials that demonstrate this behaviour are described as **ohmic conductors** and their resistance is expressed as an equation:

$$\text{resistance} \ (\Omega) = \frac{\textbf{voltage}}{\textbf{current}} \quad \frac{\text{(V)}}{\text{(A)}}$$

ISBN: 9780170195997

Expressed mathematically:

$$R = \frac{V}{I}$$

Resistance is measured in ohms (Ω), with an ohmmeter, connected across a component and it can be calculated using the formula above. We can see from the equation that $1\,\Omega = 1\,VA^{-1}$.

Worked example: Resistance

Calculate the resistance of a lamp if a current of 1.40 A flows through it when it is connected to a 9.0 V power supply.

Solution

Given	$I = 1.40$ A
	$V = 9.0$ V
Unknown	$R = ?$
Equations	$V = IR$
Substitute	$9.0 = 1.40 \times R$
Solve	$\dfrac{9.0}{1.40} = R$
	$R = 6.4\,\Omega$ (2 s.f.)

Power (*P*)

The **power rating** (*P*) of a device describes the rate at which the energy is changed by the device. It is measured in **watts** (*W*). Power can be expressed as a formula:

$$\text{power (W)} = \frac{\text{change in energy} \quad (\text{J})}{\text{time} \quad (\text{s})}$$

Expressed mathematically:

$$P = \frac{\Delta E}{t}$$

It is worth noting that in electrical circuits the power can also be expressed as:

$$P = VI$$

This can be otherwise written as:

$$\text{power} = \text{voltage} \times \text{current}$$
$$\text{(W)} \quad\quad \text{(V)} \quad\quad\quad \text{(A)}$$

ISBN: 9780170195997

Worked example: Power

Calculate the power of a starter motor that draws (takes) a current of 132 A when connected to a 12 V battery.

Solution

Given	$I = 132$ A
	$V = 12$ V
Unknown	$P = ?$
Equations	$P = IV$
Substitute	$P = 12 \times 132$
Solve	$P = 1584$ W
	$P = 1600$ W (2 s.f.)

Exercise 6B

Calculating resistance, charge and power

1 Solve the following problems using the equation $I = \dfrac{Q}{t}$.

a Calculate the current through a resistor if a charge of 54 C flows though it every 30 s.

_____

b Calculate how long it will take for 175 C to flow though a lamp when connected to a current of 0.125 A. Express your answer to the correct number of significant figures.

_____

_____

c Determine the quantity of charge that flows through a lamp when a current of 250 mA flows for five minutes.

_____

_____

d Rechargeable batteries advertise their number of 'mA h' – the larger the number means the better the battery. But what is a 'mA h'?

i Convert 2450 mA h to SI units and state an alternative unit.

_____

_____

_____

_____

ii How many electrons can be supplied by this battery? A single electron has a charge of -1.6×10^{-19} C.

_____

_____

ISBN: 9780170195997

2 Solve the following problems using the equation $V = \dfrac{\Delta E}{Q}$.

a A lamp transforms 120 J of electrical energy into light energy when 0.50 C flow through it. Calculate the voltage across the lamp.

---

b Determine the amount of energy transformed from chemical to electrical in a 12 V car battery when 1.40×10^2 C of charge is drawn from it.

---

c While recharging a laptop computer, a 240 V charger transfers 1.30 MJ of energy to the computer's battery. Calculate the number of charges which pass through the charger.

---

3 Solve the following problems using the equation **V = IR**.

a A current of 5.0 A is drawn by an electric heater when connected to a 240 V supply. Calculate the resistance of the heater.

---

b Calculate the voltage across a 15 Ω resistor when a current of 200 mA flows through it.

---

c A 12.0 V car battery is connected to a starter motor of resistance 0.83 Ω. Calculate the current that flows through the starter motor.

---

4 Solve the following problems by combining any two of $V = IR,\ I = \dfrac{Q}{t},\ V = \dfrac{\Delta E}{Q}$ or **P = IV**.

a Calculate the amount of energy transformed by a lamp connected to a 15.0 V supply for 24 s if a current of 0.20 A flows through it.

---

b 180 C of charge flows through a resistor of resistance 40 Ω in 5.0 minutes. Calculate the voltage across the resistor.

---

c An electric heater has a power rating of 2.40 kW and is connected to a 240 V supply and switched on for 2.0 hours. Calculate the amount of charge that flows through the heater in this time.

ISBN: 9780170195997

6.3 Electric circuits

The symbols below represent some of the components in an electrical circuit.

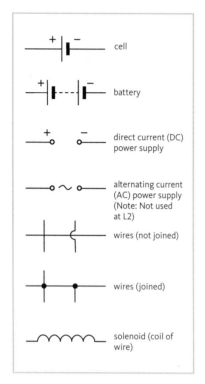

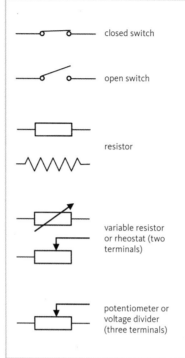

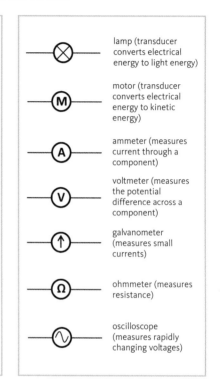

For energy to be transferred from the source to the load the circuit must provide a continuous path from the positive terminal of the energy source to the negative terminal. A complete circuit is referred to as a **closed circuit**.

In an **open circuit** the path is broken so it does not allow current to flow, and no energy can be transferred.

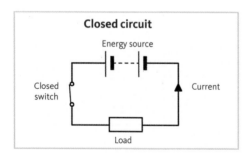

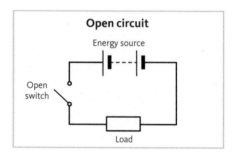

Resistors in series

When resistors are connected in series the **same current** (I) passes through each of them in turn.

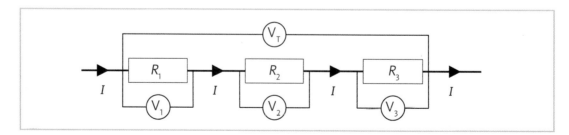

ISBN: 9780170195997

The total voltage across resistors in series is equal to the **sum of the individual voltages**. The total resistance due to resistors connected in series is equal to the sum of the individual resistances. The following series resistor formula will work for any number of resistors connected in series:

$$R_T = R_1 + R_2 + R_3$$

 ## Worked example: Equivalent resistors (series circuit)

Calculate the value of the equivalent resistor for the following circuit.

When solving circuit problems it is extremely useful to use subscripts to identify which current, voltage or resistance you are working with to avoid mistakes.

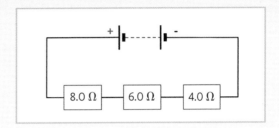

Solution

An **equivalent resistor** is the single resistor that will have the same effect as a number of resistors combined in part (or all) of a circuit. In this question the equivalent resistor represents the total resistance.

Given $R_1 = 8.0\ \Omega,\ R_2 = 6.0\ \Omega,\ R_3 = 4.0\ \Omega$

Unknown $R_T = ?$

Equations $R_T = R_1 + R_2 + R_3$

Substitute $R_T = 8.0 + 6.0 + 4.0$

Solve $R_T = 18\ \Omega$

Resistors in parallel

When resistors are connected in parallel, each branch will experience the **same potential difference (V)**.

The total current through resistors in parallel is equal to the **sum of the individual currents**. With each additional parallel branch the total resistance decreases due to the extra paths the current can take. The total resistance can be calculated using the following parallel resistor formula:

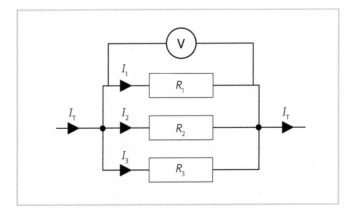

$$\frac{1}{R_T} = \frac{1}{R_1} + \frac{1}{R_2} + \frac{1}{R_3}$$

Worked example: Equivalent resistors (parallel circuit)

Determine the value of the equivalent resistor for the following circuit.

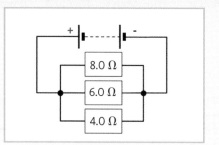

Solution

An equivalent resistor is the single resistor that will have the same effect as a number of resistors combined in part (or all) of a circuit. In this question the equivalent resistor represents the total resistance.

Given	$R_1 = 8.0\ \Omega,\ R_2 = 6.0\ \Omega,\ R_3 = 4.0\ \Omega$
Unknown	$R_T = ?$
Equations	$\dfrac{1}{R_T} = \dfrac{1}{R_1} + \dfrac{1}{R_2} + \dfrac{1}{R_3}$
Substitute	$\dfrac{1}{R_T} = \dfrac{1}{8.0} + \dfrac{1}{6.0} + \dfrac{1}{4.0} = \dfrac{13}{24}$

This value represents $\dfrac{1}{R_T}$ not R_T (The most common error when solving parallel problems is to forget to invert this value!)

Solve	$R_T = \dfrac{24}{13}\ \Omega$ or $R_T = 1.846\ \Omega$

$R_T = 1.8\ \Omega$ (2 s.f.)
Which is less than the lowest resistance branch with a resistance of 4.0 Ω.

Parallel and series circuits combined

Most circuits will involve components being connected in both series and parallel, in which case there is a strict order that each section of the circuit must be solved in to find an equivalent resistor. Consider the circuit diagram below.

STEP 1: Combine any resistors in **series** on the main circuit or in parallel branches using the resistors in series formula:

$$R_{12} = R_1 + R_2$$
$$R_{34} = R_3 + R_4$$
$$R_{567} = R_5 + R_6 + R_7$$

(If there aren't any resistors in series, skip to STEP 2.)

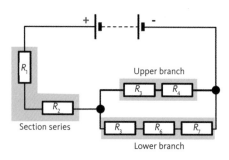

STEP 2: Combine the equivalent resistors in **parallel** using the resistors in parallel formula:

$$\frac{1}{R_{34567}} = \frac{1}{R_{34}} + \frac{1}{R_{567}} + \ldots$$

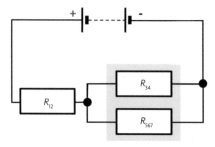

ISBN: 9780170195997

STEP 3: Combine these equivalent resistors in
series using the resistors in series formula:

$$R_T = R_{12} + R_{34567}$$

The final equivalent resistor represents the
single resistor that would have the same
effect as all the other resistors combined.

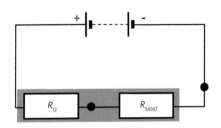

Exercise 6C

Calculating resistance and power output in series and parallel

1   Solve the following problems using the equation $R_T = R_1 + R_2 + ...$

   a   Calculate the total resistance of the series branch shown below.

   b   The series branch below has a total resistance of 92 Ω. Calculate the size of the unknown
       resistor.

   c   The series branch below is made of two identical resistors and has a total resistance of
       64 Ω. Calculate the size of the unknown resistors.

2   Solve the following problems using the equation $\frac{1}{R_T} = \frac{1}{R_1} + \frac{1}{R_2} + ...$

   a   Calculate the total resistance of the following parallel branch.

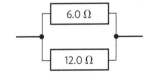

ISBN: 9780170195997

b The following parallel branch is made of three resistors and has a total resistance of 6.0 Ω. Calculate the size of the unknown resistor.

_____

_____

_____

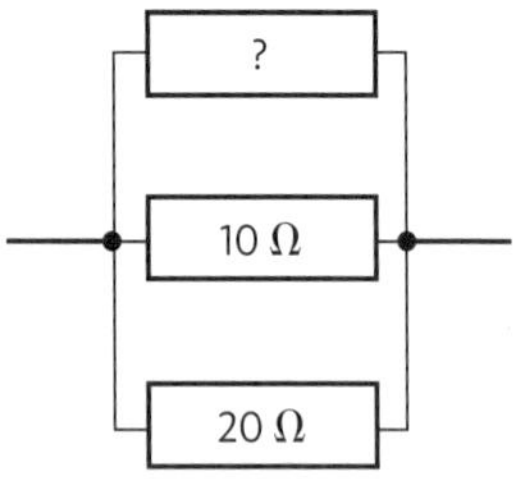

c The following parallel branch is made of two identical resistors and has a total resistance of 64 Ω. Calculate the size of the unknown resistors.

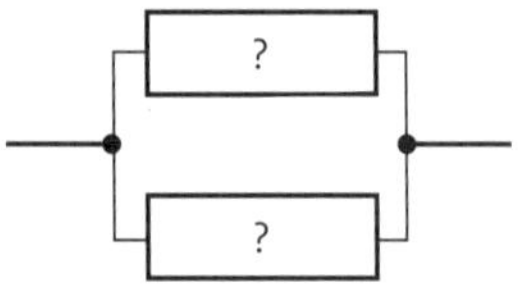

_____

_____

_____

3 Solve the following problems by combining the series and parallel equations.

a Calculate the total resistance of the resistor network shown below.

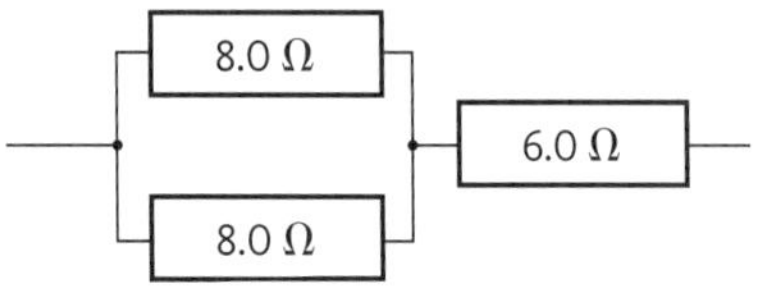

_____

_____

b Calculate the total resistance of the resistor network shown opposite.

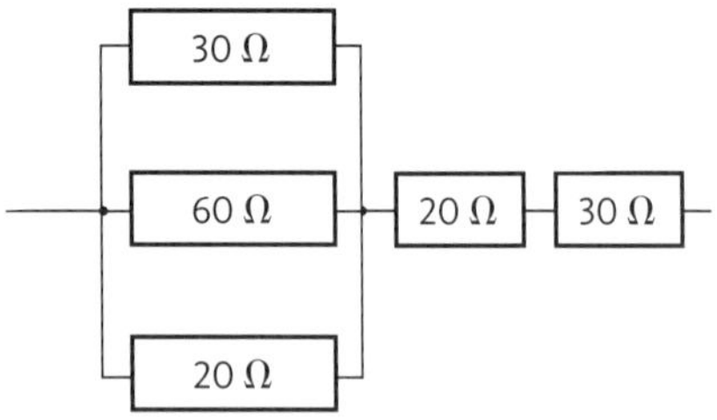

_____

_____

_____

c The resistor network below has a total resistance of 12Ω.

i Calculate the size of the unknown resistor.

_____

_____

_____

_____

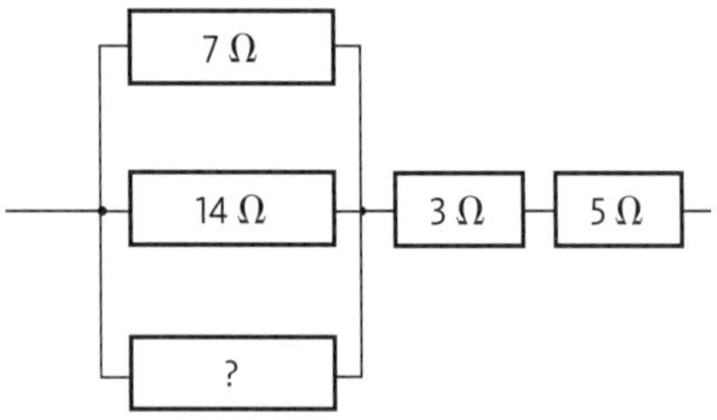

ii Explain what will happen to the total resistance if another resistor is added in **parallel** with the 7 Ω resistor.

_____

_____

_____

ISBN: 9780170195997

iii Explain what will happen to the total resistance if another resistor is added in **series** with the 5 Ω resistor.

_____

_____

_____

d The resistor network at right is made of five identical resistors and has a total resistance of 105 Ω. Calculate the size of the unknown resistors.

_____

_____

_____

_____

_____

4 When camping Anna uses a battery operated kettle, which she attaches to the battery of her campervan. The kettle has a label attached to it which reads:

Boil-o-matic
Model: FZ1K5 15 GR8T
12 V   180 W

a Use the information from the label to determine the maximum current that flows through the kettle.

_____

_____

b Show that the resistance of the heating element in the kettle is 0.80 Ω.

_____

_____

_____

c Explain why the resistance of the kettle will be lower when the heating element is cold.

_____

_____

_____

_____

_____

_____

d Calculate the amount of electrical energy Anna's kettle will transform to heat if she leaves it running for 4.0 minutes.

_____

_____

_____

_____

The campervan's power socket has a small warning light, wired in series with the sockets, which lights up whenever an appliance is plugged in. It is designed to 'blow' if the current exceeds 20 A, breaking the circuit and making it safe.

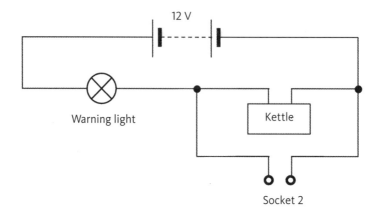

12 V

Warning light

Kettle

Socket 2

e   The total resistance of the circuit with the warning light and the kettle plugged in is 0.90 Ω. Determine the resistance of the warning light.

_____

_____

f   Determine the power output of the kettle when it is connected in the circuit above.

_____

_____

_____

_____

_____

Anna wants to plug another appliance into Socket 2 but is concerned that it will cause the warning light to blow by exceeding the maximum current of 20 A.

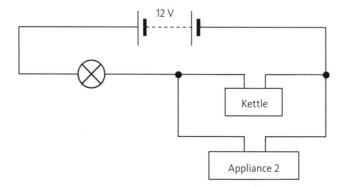

12 V

Kettle

Appliance 2

g   Show that the minimum resistance of a second appliance that could be plugged in to Socket 2 without causing the warning light to blow is 1.3 Ω.

_____

_____

_____

_____

_____

ISBN: 9780170195997

h Explain how plugging in a second appliance with this minimum resistance will affect the time taken for the kettle to boil the water.

_____

_____

_____

_____

_____

_____

5 John builds a toy car for his son Max to play with. He connects an interior lamp in series with the battery and two identical head lamps and two identical tail lamps in parallel with each other, as shown in the circuit diagram below.

Note: The head lamps are not the same as the tail lamps.

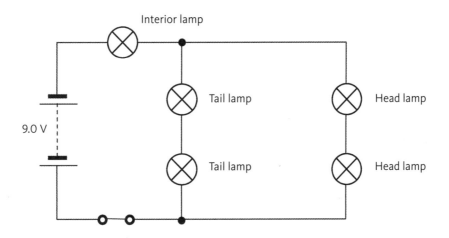

a Each head lamp has a resistance of 2.0 Ω and when operating normally a current of 1.5 A flows through them. Show that the voltage across each head lamp is 3.0 V.

_____

_____

b Explain why the current through both head lamps is the same.

_____

_____

_____

c State the voltage across both tail lamps and the interior lamp and explain your reasons.

_____

_____

_____

_____

_____

d The tail lamps are one third as bright as the head lamps. Calculate the resistance of a single tail lamp.

_____

_____

_____

_____

_____

e Show that the total resistance in the circuit is 4.5 Ω.

_____

_____

_____

_____

_____

_____

_____

f Max plays with the car for 9.0 minutes. Calculate the energy changed by the battery during this time.

_____

_____

_____

_____

_____

g The following day when Max switches on the car the head lamps have stopped working. Describe and explain the effect this will have on the other components in the circuit.

_____

_____

_____

_____

_____

_____

_____

_____

_____

ISBN: 9780170195997

6 Marilyn is a keen walker and often wears a head torch when walking at night. The torch uses four 1.5 V batteries wired in series with an 18 W lamp, as shown below.

6.0 V

18 W

a How many joules of energy will the battery supply to the lamp in 20 minutes?

_____

_____

_____

_____

_____

b Calculate the resistance of the lamp when it is switched on.

_____

_____

_____

_____

_____

The graph shows how the voltage and current are related for the lamps used in the head torch.

c Using the graph determine what would happen to the total power output if a second 18 W lamp was connected in series with the first 18 W lamp.

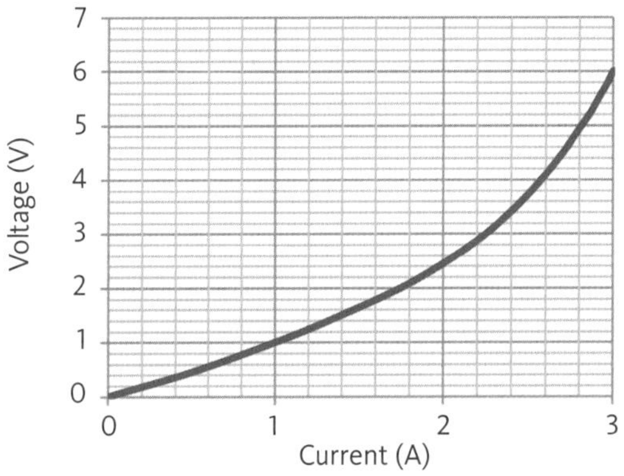

_____

_____

_____

_____

_____

_____

_____

_____

_____

_____

When Marilyn arrives at the camp site she uses a battery pack to operate her navigation computer, her mobile phone and a small fan, as shown in the diagram. The fan has a resistance of 1.20 Ω, the computer has a resistance of 8.00 Ω and the phone has a resistance of 12.00 Ω.

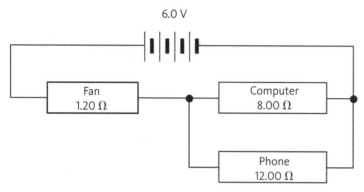

6.0 V

Fan
1.20 Ω

Computer
8.00 Ω

Phone
12.00 Ω

d Calculate the total resistance of the circuit.

_____

_____

_____

_____

_____

_____

_____

_____

e Calculate the current through the mobile phone.

_____

_____

_____

_____

_____

_____

ISBN: 9780170195997

7 William is investigating the relationship between the resistance of a conductor and its diameter. He cuts six conductors into 150 mm long pieces. Each one is made from the same material but has a different diameter. He then uses an ohmmeter to measure the resistance. The measurements are shown below.

a Identify any anomalous values and calculate the averages for the data.

Diameter (mm) (±0.1 mm division error)	Resistance (kΩ) (±0.02 kΩ division error)				
	Trial 1	Trial 2	Trial 3	Average	
1.5	39.02	39.04	39.05		
2.0	21.96	21.95	21.97		
2.5	14.05	14.06	14.06		
3.0	9.76	9.76	9.77		
3.5	7.18	7.17	7.18		
4.0	5.49	5.50	5.50		
4.5	4.34	4.33	4.34		

b Draw a graph of resistance (y) against diameter (x) using the data above, and draw on the line of best fit.

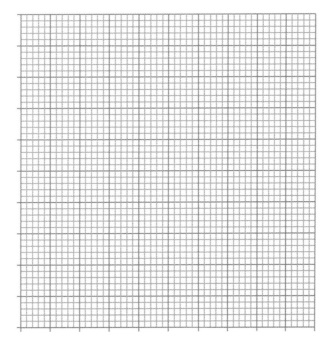

c State the relationship based upon the shape of the graph.

_____

d Complete the last column of the table by processing the data. Include an appropriate quantity and unit.

ISBN: 9780170195997

e Plot a second graph using your processed data on the graph below.

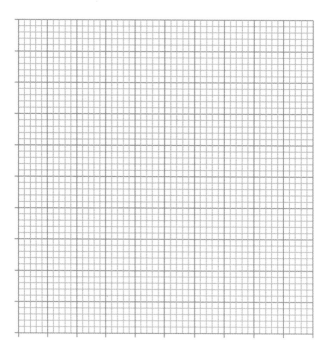

f Determine the gradient of the graph and the intercept and state the equation of the line. Provide units with all your values.

_____

_____

_____

g Compare the equation of the line to the theoretical equation $R = \dfrac{4\rho L}{\pi} \times \dfrac{1}{d^2}$ and hence determine the resistivity, ρ of the conductors.

_____

_____

_____

ISBN: 9780170195997

6.4 Electromagnetism

Uniform magnetic fields

A uniform magnetic field forms between the poles of a horseshoe magnet.

ISBN: 9780170195997

Representing uniform magnetic fields

If the magnetic field is travelling perpendicular to the page (towards or away from the reader), then physicists use a dot (•) to represent the magnetic field coming out of the paper and a cross (×) to represent the magnetic field going into the paper.

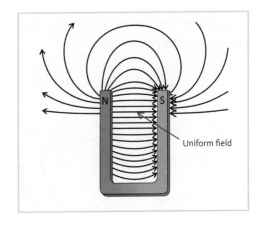

Uniform field

Magnetic field **out** of the page

The field lines are travelling towards the reader away from a north pole inside the page

The dot and cross represent the direction of the magnetic field perpendicular to the page

This notation can be remembered by thinking of a dart. The point of the arrow would look like a dot as it travelled towards you. The tail feathers would look like a cross as they travelled away from you.

Magnetic field **into** the page

The field lines are travelling away from the reader towards a south pole inside the page

Magnetic fields around a current carrying wire

When a current (I) flows through a wire then a magnetic field (B) is produced around the wire, which causes a compass needle to align itself **perpendicular** to the direction of the current.

The **strength of a magnetic field** is measured in **tesla** (T), and is dependent on:

- the size of the **current**
- the **distance** (d) from the wire
- how well the magnetic field permeates (spreads) through the medium surrounding it.

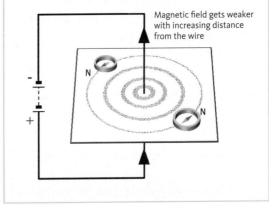

Magnetic field gets weaker with increasing distance from the wire

The direction of the magnetic field is dependent upon the direction of the current flowing through the wire.

The right-hand grip rule for wires

The direction of the field around a current-carrying wire can be determined using the **right-hand grip rule**. If the current-carrying wire is held in the right hand, then:

- the thumb points in the direction of the current
- the fingers point in the direction of the field lines.

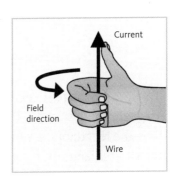

Current

Field direction

Wire

Representing currents and magnetic fields around wires

If the conductor in which the current is travelling is perpendicular to the paper that the field is being drawn on (i.e. towards or away from the reader), then physicists use a dot ⊙ to represent the current coming out of the paper and a cross $\otimes$ to represent the current going into the paper.

The field lines become further apart as the distance from the wire increases, showing that the field gets weaker with distance.

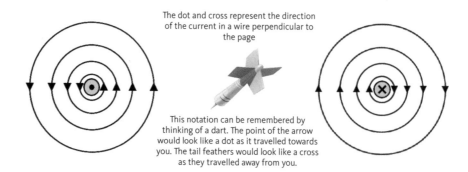

The dot and cross represent the direction of the current in a wire perpendicular to the page

This notation can be remembered by thinking of a dart. The point of the arrow would look like a dot as it travelled towards you. The tail feathers would look like a cross as they travelled away from you.

Force on a current carrying conductor in a magnetic field

A current-carrying wire is surrounded by a magnetic field, and when it is placed perpendicular to a uniform magnetic field the fields interact, resulting in the wire experiencing a force perpendicular to the direction of the magnetic fields and the current.

Where the fields act in opposite directions they cancel out, producing a weaker field. Where the fields act in the same direction they add together, producing a stronger field. The result is that the wire is pushed in the direction of the weaker field.

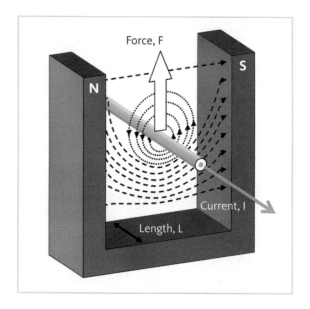

Size of the force

The size of the force depends upon the strength of the magnetic field produced by the horseshoe magnet, the size of the current and the perpendicular length of the wire **inside the field**.

The relationship can be expressed as the formula:

force = magnetic field strength × current × length
(N)   (T)   (A)   (m)

Expressed mathematically:

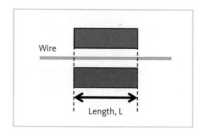

$$F = BIL_{\perp}$$

If the wire is longer than the magnetic field then the length of the magnetic field is used. If the magnetic field is longer than the wire then the length of the wire is used. (But remember: always choose the shorter length!)

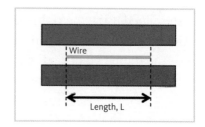

ISBN: 9780170195997

Direction of the force: Fleming's left-hand rule

The **direction** in which the force acts **on the conductor** depends upon:

- the direction of the magnetic field provided by the permanent magnet
- the direction of the current flowing through the wire.

This can be remembered using **Fleming's left-hand motor rule.**

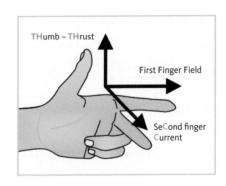

Remember

L
e
Force
t

Worked example: Force on a wire

An aluminium rod 7.0 cm long is sitting at rest on two aluminium rails, which are connected to the positive and negative terminals of a battery, as shown. The rod and rails are placed inside a uniform magnetic field of strength 2.0 T. When the power supply is turned on a current of 3.0 A flows through the rod causing it to be pushed out of the field. Calculate:

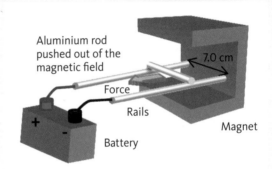

a The direction of the magnetic field, and identify which pole is north and south.

b The strength of the force on the wire.

Solution

a Using Fleming's left-hand rule, if the force on the wire is to the left and the current flows from the back rail then the field must be upwards. This means that the north pole must be at the bottom and the south pole at the top

b

Given	I = 3.0 A
	B = 2.0 T
	L = 7.0 cm = 0.070 m
Unknown	F = ?
Equations	$F = BIL$
Substitute	F = 2.0 x 3.0 x 0.070
Solve	F = 0.42 N (2 s.f.)

ISBN: 9780170195997

Exercise 6D

Conductors, current and field strength

1 Solve the following problems using **F = BIL**.

a A conductor of length 0.050 m is placed in a magnetic field of strength 2.8 T. Calculate the size of the magnetic force when a current of 1.3 A flows through the conductor.

_____

b A thin aluminium foil sheet of length 12 cm is placed in a magnetic field. When a current of 0.3 A flows through the sheet it experiences a force of 1.6 N.

   i Calculate the strength of the magnetic field.

   _____

   _____

   ii State two possible units for magnetic field strength.

   _____

   _____

c An overhead power cable carries a current of 1.50×10^2 A, and experiences a force of 1.35 N due to the Earth's magnetic field of strength 54 μT. Calculate the length of the cable.

_____

_____

_____

2 Draw labelled arrows on the following diagrams to show the direction of the magnetic field and the current in the following diagrams. Then use Fleming's left-hand rule to determine if the rod will move, and if so, in which direction.

a
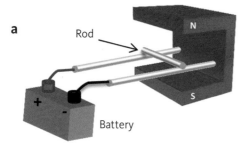

b
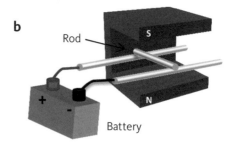

c
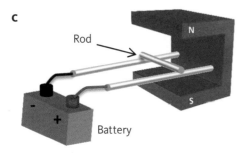

d
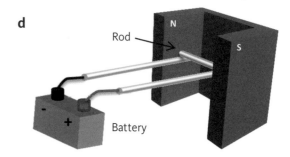

ISBN: 9780170195997

3 Solve the following problem using the equation $F = BIL$, and the correct lengths and units.

A wire 75.0 cm long is placed inside a magnetic field of strength 600 mT, as shown in the diagram. The field is 35.0 cm wide. A current of 5.0 A is switched on, resulting in a magnetic force acting on the wire.

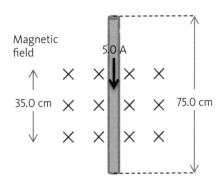

i Calculate the size of the magnetic force on the wire.

_____

_____

_____

ii Draw an arrow on the diagram to show the direction in which the magnetic force acts.

4 Solve the following problems using $F = BIL$, and $V = IR$.

A 9.0 V battery is attached to a doorbell of resistance 22.5 Ω. One of the wires supplying current to the doorbell is resting on a magnet which is 5.0 cm long. When the doorbell is switched on the wire experiences a force of 46 mN. Calculate the strength of the magnetic field.

_____

_____

_____

_____

_____

_____

5 Caleb is experimenting with his electricity and magnetism kit. He connects a 9.0 Ω lamp to a 12.0 V battery using two wires and a thin strip of aluminium foil 10 cm long. The aluminium foil is placed between the poles of an 8.00 cm long horseshoe magnet with a magnetic field of strength 1.60 T. When he switches on the circuit the foil experiences a magnetic force.

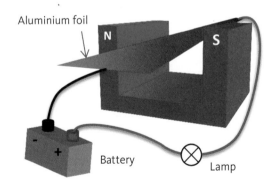

a Draw an arrow on the diagram to show the direction of the current and the direction of the magnetic field produced by the horseshoe magnet.

b Draw an arrow on the diagram to show the direction of the magnetic force on the aluminium foil.

ISBN: 9780170195997

c Explain why the foil experiences a magnetic force in the direction you have stated.

_____

_____

_____

_____

_____

d Calculate the size of the magnetic force acting on the foil. Give your answer to the correct number of significant figures.

_____

_____

_____

_____

_____

e Explain what will happen to the size of the magnetic force if the length of the aluminium foil is doubled.

_____

_____

_____

_____

_____

Force on charged particles moving in a magnetic field

When a charged particle is at rest in a magnetic field there is no force acting upon it. But when the charged particle is moving in a uniform magnetic field it experiences a force perpendicular to its direction of motion.

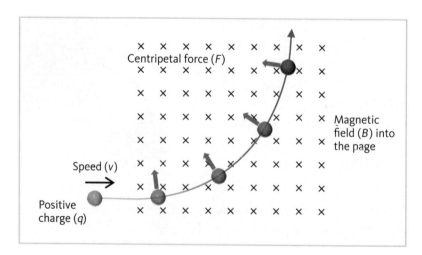

As the force acts perpendicular to the direction of motion, the charged particle will move along a **circular path**, and will leave the magnetic field at the **same speed** that it entered (see page 130 on Circular motion).

ISBN: 9780170195997

Size of the force

The **size of the force** depends upon the strength of the magnetic field (B), the size of the charge (q) and the speed (v) at which the particle enters. The relationship can be expressed as a formula:

force = magnetic field strength × charge × speed
(N)   (T)   (C)   (m s^{-1})

Expressed mathematically:

$$F = Bqv_{\perp}$$

Direction of the force: Fleming's left-hand Rule

The direction of the force on a **positively** charged particle can be found using Fleming's left-hand rule (see page 217).

Negative charged particles are always **deflected in the opposite direction** to positive charged particles.

Remember that **current is the flow of positive charges** so the second finger should point:

- in the same direction as the motion of a positive charge
- in the opposite direction to the motion of a negative charge.

Worked example: Force on a charged particle

An electron of charge -1.602×10^{-19} C is travelling at a speed of 2.00×10^6 m s^{-1} through a magnetic field of strength 3.40 T. Determine the size of the magnetic force acting on the electron.

Solution

Given

$q = -1.602 \times 10^{-19}$ C
$B = 3.40$ T
$v = 2.00 \times 10^6$ m s^{-1}

Unknown $F = ?$

Equations $F = Bqv$

Substitute $F = 3.40 \times (-1.602 \times 10^{-19}) \times 2.00 \times 10^6$

Solve $F = -1.09 \times 10^{-12}$ N (3 s.f.)

The negative sign tells us that the force acts in the opposite direction to the magnetic field.

ISBN: 9780170195997

Exercise 6E

Magnetism and radiation

1 Solve the following problems using **F = Bqv**.

a A hydrogen ion with a charge of 1.6×10^{-19} C experiences a force of 2.3×10^{-16} N when it passes through a magnetic field of strength 2.5 T. Calculate the speed of the ion.

_____

_____

b An oxygen ion has a charge of 3.2×10^{-19} C and is moving at 482 m s^{-1} in a magnetic field of strength 650 mT. Calculate the size of the force acting on the oxygen ion.

_____

_____

c An unknown ion is travelling at 2.7×10^{8} m s^{-1} when it enters a magnetic field of strength 4.9 T. Once in the field the particle experiences a force of 6.35×10^{-10} N. Calculate the charge on the particle and then determine how many electrons are missing from the ion.

_____

_____

_____

_____

2 Using Fleming's left-hand rule, draw an arrow on the diagrams below to show the direction of the force (if any) acting on the charged particles.

   a   **b**   **c**   **d**

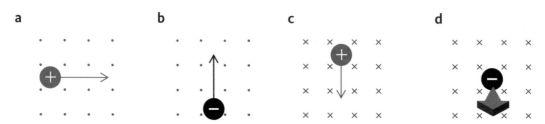

3 Marie is studying the radiation given off by a radioactive rock. She places the rock in a lead box with a small hole at one end to produce a beam of radiation. She passes the beam through a magnetic field of strength 5.6 T and uses a detector to reveal the paths the radiation takes. The diagram below shows her results.

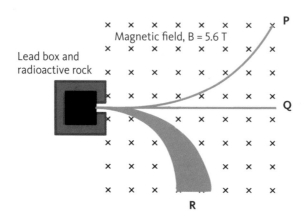

ISBN: 9780170195997

Marie discovers that three types of radiation **P**, **Q**, and **R**, are emitted by the rock.

a   Determine the charge on each type of radiation by considering the direction in which they travel inside the magnetic field.

    i   Radiation **P**: _____

    ii Radiation **Q**: _____

    iii Radiation **R**: _____

b   Explain clearly how you determined the charge on each of the types of radiation.

_____

_____

_____

_____

_____

_____

_____

_____

All the particles arriving at R have the **same mass and charge** but Marie notices that they gradually spread out and don't all travel along the same **semi-circular** path like radiation P.

c   Explain why the radiation arriving at R spreads out.

_____

_____

_____

_____

_____

_____

_____

By measuring the radius of the path taken by the radiation P, Marie calculates that the force acting on it is 2.688×10^{-11} N. Marie also finds the size of the charge on radiation P is 3.2×10^{-19} C.

d   Calculate the speed of radiation P.

_____

_____

_____

_____

_____

4 A mass spectrometer is a device used by scientists to determine the different elements present in a chemical sample. The sample is ionised by removing electrons and then the positive ions are accelerated towards a magnetic field of strength 0.20 T. The diagram at right shows the path of two positively charged helium ions travelling through the magnetic field.

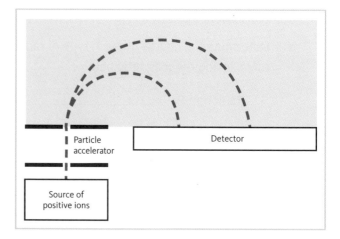

a By considering the direction in which the positive ions move, state whether the magnetic field is acting into the page or out of the page.

_____

A singly charged helium ion (He^{1+}), of charge 1.6×10^{-19} C, enters the magnetic field at a speed of 1.50×10^5 m s^{-1}.

b Calculate the size of the force acting on the He^{1+} ion.

_____

_____

c Explain what will happen to the speed of the ion as it travels through the magnetic field.

_____

_____

_____

_____

The two dotted lines in the diagram represent the path of a singly charged helium ion (He^{1+}), and a doubly charged helium ion (He^{2+}).

d Identify on the diagram which path belongs to the He^{2+} ion and explain your reasons for choosing that path.

_____

_____

_____

_____

_____

_____

_____

_____

ISBN: 9780170195997

Induced current

If a conductor, which is connected to sensitive ammeter called a **galvanometer**, is moved perpendicular to a magnetic field then a small current is induced in the wire.

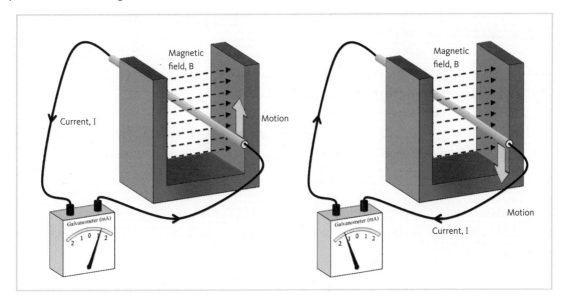

In the example above:

- when the conductor moves upwards the current flows anticlockwise and the needle flicks to the right.
- when the conductor moves down the current flows clockwise and the needle flicks to the left.

The effect is similar if the magnet is moved instead of the conductor. It is the **relative motion** of the conductor to the magnet that is important. However, **no current is induced** when the conductor is:

- stationary
- moved in line with the magnetic field.

Direction of the current: Fleming's right-hand rule for Induction

The direction in which the induced **current** travels around the circuit depends upon:

- the direction of the **magnetic field** provided by the permanent magnet
- the direction the **conductor is moving relative to the magnetic field**.

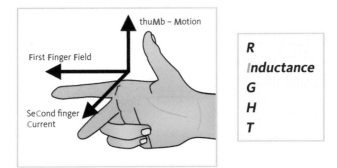

This can be remembered using Fleming's right-hand rule for induction. *Remember: the seCond finger refers to the direction of the Current, i.e. positive charge flow. This means that the electrons flow in the opposite direction.*

Faraday's Law and induction

When a conductor is pushed through a magnetic field the electrons in the conductor experience a force, causing them to move along the conductor towards one end. The end that gains electrons becomes negatively charged and the end which loses electrons becomes positively charged, **inducing a voltage** across the ends of the conductor.

Faraday found that the induced voltage increased if:
- the **magnetic field strength** (B) increased
- the **speed** (v) of the relative motion increased
- the **length of the conductor** (L) in the magnetic field increased. This could be achieved by either:
  - looping the conductor so that it passed through the field several times
  - or making the magnets longer.

These observations led Faraday to conclude that the size of the induced voltage is directly proportional to the rate at which the conductor cuts magnetic field lines. Therefore **Faraday's Law** can be written as:

voltage = magnetic field strength × speed × length
(V)   (T)   (m s⁻¹)   (m)

Expressed mathematically:

$$V = BvL_\perp$$

Lenz's Law

When a conductor is pushed through a uniform magnetic field it induces a current in the conductor. The induced current will have a magnetic field around it, which interacts with the uniform magnetic field, producing a force on the conductor that opposes the original pushing force.

Worked example: Flying through the Earth's magnetic field

A Boeing 747 has a wingspan of 64.4 m and flies at a cruising speed of 250 m s⁻¹. Calculate the size of the voltage that is induced across the wing tips of the aircraft if it is flying through the Earth's magnetic field of strength 52 µT.

64.4 m

Solution

Given	$L = 64.4$ m
	$v = 250$ m s⁻¹
	$B = 52 \times 10^{-6}$ T
Unknown	$V = ?$
Equations	$V = BvL$
Substitute	$V = 52 \times 10^{-6} \times 250 \times 64.4$
Solve	$V = 0.84$ V (2 s.f.)

ISBN: 9780170195997

Exercise 6F

Induction and the laws of magnetism

1 Solve the following problems using **V = BvL**.

a A wire of length 0.20 m moves through a magnetic field of strength 1.3 T at a speed of 1.9 m s⁻¹. Calculate the voltage induced across the ends of the wire.

_____

b A voltage of 29.0 mV is induced across the ends of the wings of a small plane with a wingspan of 8.28 m, as it flies perpendicular to the Earth's magnetic field of strength 52 µT. Calculate the speed of the plane.

_____

_____

c A skydiver carrying a copper rod falls through the Earth's magnetic field (of strength 52 µT.) The skydiver quickly reaches terminal velocity of 200 km h⁻¹, at which speed a voltage of 2.0 mV is induced across the ends of the rod. Calculate the length of the rod.

_____

_____

_____

2 Using Fleming's right-hand rule, draw an arrow on the diagrams below to show the direction of the induced current in each rod and identify the end that becomes positive.

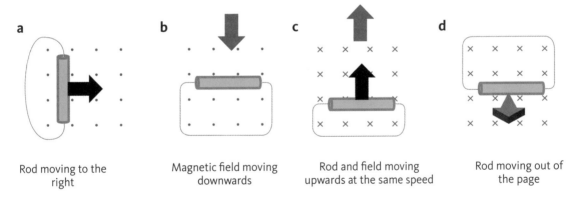

| a | b | c | d |
| Rod moving to the right | Magnetic field moving downwards | Rod and field moving upwards at the same speed | Rod moving out of the page |

3 Caleb connects an aluminium foil to a sensitive ammeter and moves the wire foil up and down between the poles of an 8.00 cm long horseshoe magnet with a magnetic field of strength 1.60 T. The components in the circuit provide a resistance of 0.5 Ω.

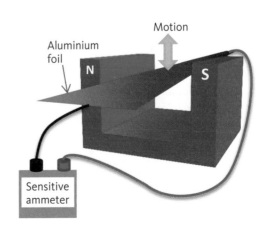

a Label the diagram with an arrow to show the direction of the current flow in the foil when it is moving upwards.

ISBN: 9780170195997

b Calculate the size of the induced current if the foil moves at maximum speed of 6.0 m s^{-1} through the field.

_____

_____

c Explain clearly in terms of the movement of electric charges, what happens as the foil starts moving upwards through the magnetic field.

_____

_____

_____

_____

4 An aluminium rod, 0.8 m long, is pushed along two metal rails at a steady speed of 7.0 m s^{-1}, perpendicular to a vertical magnet field of strength 0.50 T, as shown below. The two rails are joined together by a wire and a 1.60 Ω resistor.

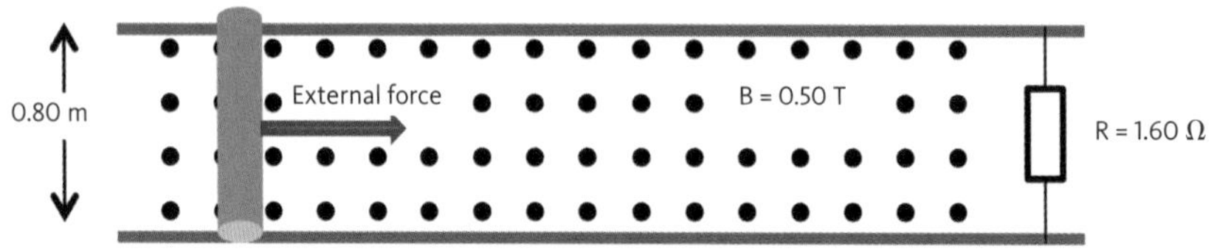

a Calculate the size of the current induced in the resistor.

_____

_____

To keep the rod moving east at a steady speed a constant external force must be applied to the wheels.

b Explain why the rod requires a constant force to maintain a steady speed.

_____

_____

_____

_____

_____

c Calculate the size of the force required to keep it moving at a steady speed.

_____

_____

d Calculate the amount of work done on the rod when it moves 5.0 m.

_____

_____

e Explain what happens to the energy changed when the work is done moving the rod.

_____

_____

_____

ISBN: 9780170195997

Appendices

Appendix 1: Relevant mathematics

Pythagoras' theorem

$h^2 = o^2 + a^2$

Trigonometric relations

$\sin\Theta = \dfrac{o}{h}$ $\quad$ $\cos\Theta = \dfrac{a}{h}$ $\quad$ $\tan\Theta = \dfrac{o}{a}$

$\tan\Theta = \dfrac{\sin\Theta}{\cos\Theta}$

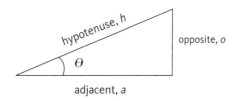

Quadratic solver

If $ax^2 + bx + c = 0$ then $x = \dfrac{-b \pm \sqrt{b^2 - 4ac}}{2a}$

Perimeters, areas and volumes

Area of a triangle	$A = \frac{1}{2} \times b \times h$
Area of trapezium	$A = \left(\dfrac{s + l}{2}\right) \times h$
Circumference of a circle	$C = 2\pi r$
Area of a circle	$A = \pi r^2$
Surface area of a sphere	$A = 4\pi r^2$
Volume of a sphere	$V = \dfrac{4}{3}\pi r^3$

Indices

$x^{-a} = \dfrac{1}{x^a}$

$x^{1/b} = \sqrt[b]{x}$

$x^c \times x^d = x^{(c+d)}$

$x^e \div x^f = x^{(e-f)}$

$(x^m)^n = x^{mn}$

Logarithms

Logarithms to base 10 are called common logarithms and are represented by 'log' or '$\log_{10}$'.
If $\log x = y$ then $x = 10^y$. Three useful rules for dealing with logarithms:

$\log PQ = \log P + \log Q$ $\qquad$ $\log \dfrac{R}{S} = \log R - \log S$ $\qquad$ $\log X^n = n \log X$

ISBN: 9780170195997

Appendix 2: Graphing with calculators (regression)

The steps below explain how a Casio graphics calculator can be used to analyse data produced during an experiment; however the process is similar on other graphics calculators. The data from page 27 will be used as an example.

STEP 1: Create a results table

Switch on the calculator and enter **STAT** mode (option 2). Enter the x-axis (usually independent) data into **List 1** and the y-axis (usually dependent) data into **List 2**.

	List 1	List 2	List 3	List 4
1	0.63	0.2		
2	1	0.5		
3	1.26	0.8		
4	1.52	1.1		
5	1.7	1.4		

| GRPH | CALC | REST | INTR | DIST | ▷ |

STEP 2: Create the graph and the equation of the line

Press **GRPH** (function key F1) and select **GPH1** (function key F1) to create a graph.

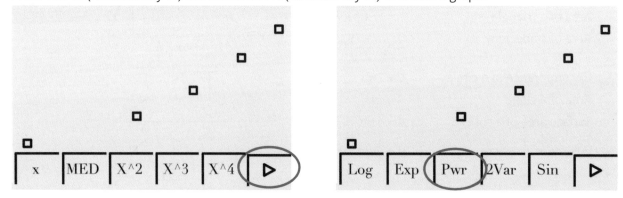

| x | MED | X^2 | X^3 | X^4 | ▷ |

| Log | Exp | Pwr | 2Var | Sin | ▷ |

If the graph is curved then: move right through the **menu** (function key F6) and select **Pwr** (function key F3) to analyse the data as a **power graph**.

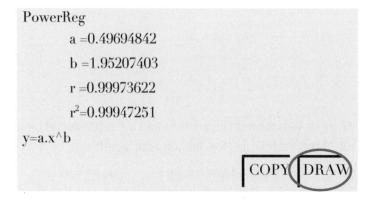

PowerReg

$a = 0.49694842$

$b = 1.95207403$

$r = 0.99973622$

$r^2 = 0.99947251$

$y = a.x^b$

| COPY | DRAW |

ISBN: 9780170195997

The r values describe how close the data points are to the line of best fit. Perfect data will produce r = 1 or -1. No relationship exists if r = 0.

The equation of the line is given in the form: $y = ax^b$. Substituting in the numbers from the calculator and replacing **x** and **y** with in **t** and **d**, gives:

$$d = 0.49\ t^{1.952}$$

The shape of the mathematical relationship can be seen by pressing **DRAW**.

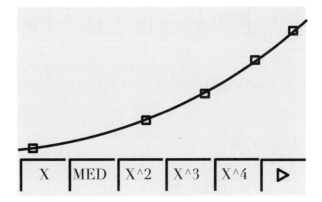

This is the line of best fit using the raw data and indicates that the relationship is likely to be a squared relationship as:

$$1.952 \approx 2$$

So the data should be processed by squaring all the time values.

STEP 3: Processing the data

Press **EXIT** and return to the List screen.
Square the x-axis data and enter it into List 1.

	List 1	List 2	List 3	List 4
1	0.37	0.2		
2	1	0.5		
3	1.59	0.8		
4	2.31	1.1		
5	2.89	1.4		

GPH1　GPH2　GPH3　SEL　　　SET

Press **GRPH** (function key F1) and select **GPH1** (function key F1) to create the new graph.

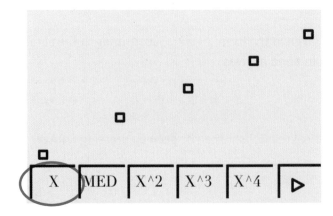

The processed graph should now be straight: select **x** (function key F1) to analyse the data as a **linear graph**.

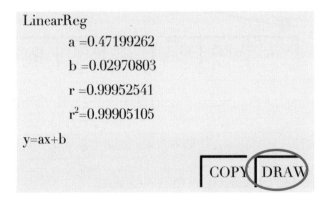

The equation of the line is given in the form: $y = ax + b$. Substituting the numbers from the calculator and replacing **x** and **y** with t^2 and **d** gives:

$$d = 0.47\ t^2 + 0.030$$

The shape of the mathematical relationship can be seen by pressing **DRAW**.

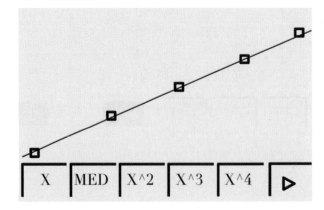

The slight difference between the Power regression and the Linear regression equations is due to the data being processed as a squared relationship i.e. t^2 not $t^{1.952}$.

ISBN: 9780170195997

Appendix 3

Periodic Table of the Elements

Atomic Number

1		
H		Relative Atomic Mass
1.0		

Main table (group numbers across the top):

1	2	3	4	5	6	7	8	9	10	11	12	13	14	15	16	17	18
3 **Li** 6.9	4 **Be** 9.0											5 **B** 10.8	6 **C** 12.0	7 **N** 14.0	8 **O** 16.0	9 **F** 19.0	10 **Ne** 20.2
11 **Na** 23.0	12 **Mg** 24.3											13 **Al** 27.0	14 **Si** 28.1	15 **P** 31.0	16 **S** 32.1	17 **Cl** 35.5	18 **Ar** 40.0
19 **K** 39.1	20 **Ca** 40.1	21 **Sc** 45.0	22 **Ti** 47.9	23 **V** 50.9	24 **Cr** 52.0	25 **Mn** 54.9	26 **Fe** 55.9	27 **Co** 58.9	28 **Ni** 58.7	29 **Cu** 63.5	30 **Zn** 65.4	31 **Ga** 69.7	32 **Ge** 72.6	33 **As** 74.9	34 **Se** 79.0	35 **Br** 79.9	36 **Kr** 83.8
37 **Rb** 85.5	38 **Sr** 87.6	39 **Y** 88.9	40 **Zr** 91.2	41 **Nb** 92.9	42 **Mo** 95.9	43 **Tc** 98.9	44 **Ru** 101	45 **Rh** 103	46 **Pd** 106	47 **Ag** 108	48 **Cd** 112	49 **In** 115	50 **Sn** 119	51 **Sb** 122	52 **Te** 128	53 **I** 127	54 **Xe** 131
55 **Cs** 133	56 **Ba** 137	71 **Lu** 175	72 **Hf** 179	73 **Ta** 181	74 **W** 184	75 **Re** 186	76 **Os** 190	77 **Ir** 192	78 **Pt** 195	79 **Au** 197	80 **Hg** 201	81 **Tl** 204	82 **Pb** 207	83 **Bi** 209	84 **Po** 210	85 **At** 210	86 **Rn** 222
87 **Fr** 223	88 **Ra** 226	103 **Lr** 262	104 **Rf** 261	105 **Db** 262	106 **Sg** 263	107 **Bh** 264	108 **Hs** 265	109 **Mt** 268	110 **Ds** 271	111 **Rg** 272	112 **Cn** (277)						

Top right: 1 **H** 1.0 ; 2 **He** 4.0

Lanthanide Series

57 **La** 139	58 **Ce** 140	59 **Pr** 141	60 **Nd** 144	61 **Pm** 147	62 **Sm** 150	63 **Eu** 152	64 **Gd** 157	65 **Tb** 159	66 **Dy** 163	67 **Ho** 165	68 **Er** 167	69 **Tm** 169	70 **Yb** 173

Actinide Series

89 **Ac** 227	90 **Th** 232	91 **Pa** 231	92 **U** 238	93 **Np** 237	94 **Pu** 239	95 **Am** 241	96 **Cm** 244	97 **Bk** 249	98 **Cf** 251	99 **Es** 252	100 **Fm** 257	101 **Md** 258	102 **No** 259

ISBN: 9780170195997

Answers

Fully worked solutions are available the accompanying Physics 2 Teacher Resource, but where possible short answers have been provided here for convenience. It is recommended that the reader print the fully worked solutions from the resource so that they have them available when attempting the exercises.

CHAPTER 1

Exercise 1A

1 a m s^{-2} acceleration
   b kg m s^{-2} force
   c s time
   d V voltage
2 a 62×10^{-9} m
   b 240×10^3 V
   c 9.9×10^9 W
   d 1.47×10^0 m
   e 583×10^{-6} kg
3 a Acceleration is equal to the rate of change of velocity.
   b Power is proportional to voltage.
   c The total current is equal to the sum of the individual currents.
4 a $k \equiv 2 \times 10^{-7}$ T m A^{-1}
   b $F_{forward} \neq F_{backward}$
   c $f \propto \frac{1}{T}$
5 a 6×10^{-5}
   b 2×10^5
   c 1.010
6 a 5×10^4 taking 100 as 1 s.f. or 4.53×10^4 taking 100 as 3 s.f.
   b 2.0×10^2 as 1.0×10^3 is 2 s.f.
   c 47.0 as 16.0 is to only 1 d.p.
7 a $\frac{3600}{1000}$ km h^{-1} = 3.6 km h^{-1}
   c 13.9 ms^{-1}.
   d 1.08×10^9 km h^{-1}.
8 a 13.0 cm
   b 6.0 cm
   c 15.0 cm
9 a $\Theta = 67°$, $\varphi = 23°$
   b $\Theta = 53°$, $\varphi = 37°$
   c $\Theta = 28°$, $\varphi = 62°$

10

a. (3 s.f.)	b. (2 s.f.)	c. (4 s.f.)
d/2 = 3.18 m	C/2π = 7.0 m	√A/π = 4.000 m
2πr = 20.0 m	44 m	2πr = 25.13 m
πr² = 31.8 m²	πr² = 150 m²	50.27 m²

11 a 18 m²
    b 12 m²
    c 300 m² (3 s.f.)
12 a 0.03
    b 16
    c 0.11
    d 10,000,000
    e 1000,000,000,000
    f 100
    g 10
    h 0.1
    i 60,466,176
    j 4,096
    k 9
13 a 1
    b 3
    c 0.477
    d -1
    e -0.097
    f 1.40
    g 1.43

CHAPTER 2

Exercise 2A

1 a Division error
      Counting error
      Parallax error
      Irregularities
      Calibration error
      Take repeat readings and calculate the averages.
   b Reaction time error
      Counting error
      Calibration error
      Take repeat readings and calculate the averages.
   c Division error
      Parallax error
      Irregularities
      Calibration error
      Take repeat readings and calculate the averages.
   d Reaction time error
      Parallax error
      Zero error
      Calibration error
      Take repeat readings and calculate the averages.
2 13 ± 1 mm.
3 1.0 ± 0.1 s.

Exercise 2B

1 a Trial 3: 0.32 N
   c Force is proportional to current
   d $m = 0.22$ N A^{-1}
      $c = 0$ N
      $F = 0.22I$
      (N) (NA^{-1}) (A)
   e $B = 2.2$ N A^{-1} m^{-1}
2 a Trial 2: 49 g and Trial 3: 76 g
   c Mass is proportional to volume
   d $m = 1.5$ g cm^{-3}
      $c = 20$ g
      $m = 1.5\ V + 20$
      (g) (g cm^{-3}) (cm³) (g)
   e Zero error
   f 1500 kg m^{-3}

Exerice 2C

1 Kristina and the glider.
   a Trial 1: 23.4 and 5.7 ms^{-1}, and
      Trial 2: 13.0 ms^{-1}.
   c Speed is inversely proportional to mass.
   f $m = 1.21$ kg ms$^{2-1}$
      $c = 0$ ms^{-1}
      $\frac{v}{(ms^{-1})} = \frac{1.21}{m}\ \frac{(kg\ ms^{-1})}{(kg)}$
   g $p = 1.21$ kg ms^{-1}
2 Jake and circular motion.
   a Trial 1: 7.1 N, and Trial 3: 2.5 N.
   c Force is proportional to speed squared.
   f $m = 0.063$ N m^{-2}s^2
      $c = 0$ N
      $F = 0.063\ v^2$
      (N) (Nm^{-2}s^2) (m^{2}s^{-2})
   g $m = 0.050$ kg
3 Arini and an oscillating mass.
   a Trial 1: 5.3s
   c Time is proportional to the square root of mass.

ISBN: 9780170195997

f $m = 0.90$ s kg$^{-\frac{1}{2}}$
   $c = 0$ s
   $T = 0.91 \sqrt{m}$
   (s) (s kg$^{-\frac{1}{2}}$) (kg$^{\frac{1}{2}}$)
g $k = 48$ kg s^{-2}.
4 Tama and the electric field strength.
   a Trial 1: 4.38 Vm^{-1}, Trial 2: 1.18 Vm^{-1}.
   c Electric field strength is inversely proportional to the distance squared.
   e $m = 0.045$ V m
      $c = 0$ V m^{-1}
      $\underset{(V\,m^{-1})}{E} = \underset{(m^2)}{\frac{0.045}{d^2}}$ (V m)
   f $q = 5.0 \times 10^{-12}$ C

Exercise 2D
Report 1: Achieved
Report 2: Not achieved
1 **a** Trial 1: 1.24 s, Trial 3: 1.79 s.
   c Distance is proportional to the time squared.
   f $m = 0.10$ m s^{-2}
      $c = 0$ m
      $d = 0.10$ t^2
      (m) (ms^{-2}) (s^2)
   g $a = 0.20$ ms^{-2}
   h 0.011 N
   i $F = -0.97$ N

CHAPTER 3
Exercise 3A
1 **a** Real, Diminished ($h_i = 1.1$ cm), Inverted, $m = \frac{1.1}{2.0} = 0.55$
   b Virtual, Diminished ($h_i = 0.5$ cm), Upright, $m = \frac{0.5}{2.0} = 0.25$
   c Virtual, Magnified ($h_i = 1.25$ cm), Upright, $m = \frac{1.25}{1.0} = 1.25$
   d Real, f = 2.0 cm, r = 4.0 cm.
   e Virtual, f = -4.0 cm, r = -8.0 cm.
   f $h_o = 1.0$ cm, $d_o = 2.0$ cm
2 **a** f = 24 cm
   b $d_i = -2.9$ m
   c $d_o = 255$ m
3 **a** $d_i = 31.5$ cm, m = 0.75, $h_i = 6.75$ cm
   b m = 0.0656, $d_i = 0.85$ m, f = 0.80 m
4 **a** f = - 5.1 m, $d_i = -1.275$ m, m = 0.75, $h_i = 0.63$ m
   b $d_o = 8.8$ m, m = 0.27, $h_o = 1.50$ m
5 $d_i + d_o = 48$ cm
6 $d_o = 22.5$ cm or 7.5 cm.

Exercise 3B
2 **a** Velocity decreases, no change in angle as the ray travels along the normal.
   b Velocity decreases, angle decreases.
   c Optical density decreases, angle increases.
   d Velocity decreases, refractive index increases.
3 **a** $n_l = 1.31$
   b $n_g = 1.6$
   c $v_w = 2.3 \times 10^8$ ms^{-1}
   d $\Delta v = -2.24 \times 10^8$ ms^{-1}
   e $\Theta_a = 44°$, $\Theta_{dev} = 16°$
4 **a** $\Theta_{30} = 42°$, $\Theta_{40} = 59°$, Θ_{50} = math error – greater than critical angle of 48.8°.
5 **b** $v_p = 2.03 \times 10^8$ ms^{-1}
   c $\Theta_g = 25°$

Exercise 3C
1 **a** Real, Magnified ($h_i = 1.8$ cm), Inverted, m = 1.2
   b Virtual, Diminished ($h_i = 1.5$ cm), Upright, m = 0.75
   c Real, Same size ($h_i = 2.5$ cm), Inverted, m = 1.0
   d Virtual, f = - 3.5 cm.
   e Virtual, f = 2.1 cm.
   f $h_o = 2.0$ cm, $d_o = 6.0$ cm,
2 **a** f = 26 cm
   b $d_o = 13$ cm

c $d_i = - 0.40$ m
3 **a** $d_i = 54$ m, m = 73, $h_i = 58$ cm
   b m = 0.005, $d_i = 11.5$ m, f = 11.4 m
4 **a** $d_i = -3.0$ m, m = 0.75, $h_i = 1.3$ m
   b $d_o = 22$ cm, m = 1.4, $h_o = 9.0$ cm
5 $d_o = 42$ cm or 28 cm.
6 $d_i + d_o = 96$ cm
8 Sue's lens experiment
   d m = −1.0 (no unit)
      $c = 4$ m^{-1}
      $\frac{1}{d_i} = \frac{-1}{d_o} + 4$
   e $f = 0.25$ m

Exercise 3D
2 A = 1.5 m
3 **a** $\lambda = 12$ m
   b v = 2.4 ms^{-1}
   c f = 0.2 Hz or s^{-1}
4 **b** T = 0.00273 s
   c $\lambda = 0.75$ m
5 **a** f = 5.5×10^{14} Hz
   b T = 1.8×10^{-15} s

Exercise 3E
1 **a** Diffraction

Exercise 3G
1 **a** Diffraction
   b **i** Boat A will experience calm water.
      ii Boat B will experience doubly large crests and troughs.
2 **a** The bright fringes are called antinodes.
   b To produce a clear pattern the waves coming from the two slits must be coherent (similar wavelength, amplitude and in phase).
   c Increasing the slit separation will cause the fringes to get closer together.
   d Increasing the frequency will cause the fringes to get closer together.
3 **e** m = 600 m^2
      $c = 0$ m
      (m) x = $\frac{600\,(m^2)}{d\,(m)}$
   f Hence $\lambda = 0.30$ m

Exercise 3I
1 **b** $f = \frac{3.0}{12} = 0.25$ Hz
   c $\lambda = \frac{1.5}{0.25} = 6.0$ m
2 **c** $f = \frac{30}{60} = 0.5$ Hz
   d $\lambda_A = \frac{0.75}{0.5} = 1.5$ m
      $\Theta_B = 70.5$ to the normal
4 **a** $\lambda_H = \frac{2.0 \times 3.0}{3.0} = 2.0$ m

CHAPTER 4
Exercise 4A
1 **a** $\underline{F} = 30$ N (E)
   b $\underline{d} = 30$ m (W)
   c $\underline{v} = 1.0$ m s^{-1} (S)
2 **a** $\underline{F} = 2000$ N (37° S of E)
   b $\underline{d} = 130$ km (23° N of W)
   c $\underline{v} = 100$ km h^{-1} (E)
3 **a** $\Delta F = 320$ N (W)
   b $\Delta d = 100$ km (W)
   c $\Delta v = 16$ m s^{-1} (W)
4 **a** $\Delta v = 6.7$ m s^{-1} (27° N of W)
   b $\Delta v = 51$ km h^{-1} (51° W of N)
5 **a** Horizontal: 153 N
      Vertically: 129 N
   b Horizontal: 16 m s^{-1}
      Vertically: 9.0 m s^{-1}
6 **a** **i** $F_{\parallel} = 153$ N
         $F_{\perp} = 867$ N

    ii F_{output} = 153 N up the slope.

  b **i** $F_{\parallel}$ = 411 N

      $F_{\perp}$ = 490 N

    ii $F_{support}$ = 490 N perpendicular

Exercise 4B

5 **a** F = 135 N

  b m = 71 kg

  c a = 6 m s^{-2}

6 **a** a = −9.00 ms^{-2}

     m = 5.0 x 10^{-4} kg

  b a = 2.0 ms^{-2}, Δv = 20 ms^{-1}

7 **a** F_{net} = 760 N

     $F_{friction}$ = 740 N

  b F_{net} = 600 N

     m = 300 kg

     a = 2.0 ms^{-2}

9 **b** F_{net} = 78 N

     F_g = 117.6 N

     $F_{friction}$ = 39.6 N

Exercise 4C

1 p = 2.46 kg m s^{-1} (left)

2 v = 13m s^{-1} (right)

3 **a** Δp = 4.48 kg m s^{-1}

     F = 2.2 kN

4 **a** Δp = 1.056 kg m s^{-1}

  b F = 176 N

Exercise 4D

2 **b** p_i = −0.16 kg m s^{-1}

  c v_F = 4.0 m s^{-1} (left) (2 s.f.)

3 **a** Δp_A = −400 kg m s^{-1}

  b N s

  c Δp_B = 400 kg m s^{-1}

  d v_F = 11 m s^{-1} (right) (2 s.f.)

4 **a** p_i = 6000 kg m s^{-1}

  b v_F = 0.55 m s^{-1} (right) (2 s.f.)

  c Δp_R = − 5455 kg m s^{-1}

5 v_{RF} = 30 m s^{-1} (right) (1 s.f.)

Exercise 4E

1 v_{ave} = 2.3 m s^{-1} (2 s.f.)

2 t = 26.5 s

3 **a** d =2.676 x 10^6 m (W)

4 **a** v =5.83 m s^{-1} (3 s.f.)

  b v =10.2 km h^{-1} (w) (3 s.f.)

5 **a** v =11 m s^{-1} (N) (2 s.f.)

  b d = 87 m (N) (2 s.f.)

  c Δv = 5.8 m s^{-1} (N) (2 s.f)

  d a = 0.73 m s^{-2} (N) (2 s.f.)

6 a = 3.3 m s^{-2} (S) (2 s.f.)

7 **a** a = 1900 m s^{-2} (←) (2 s.f.)

8 **a** a = 0.89 m s^{-2} (E) (2 s.f.)

  b d = 36 m (E) (2 s.f.)

9 **a** a = −3.8 m s^{-2} (2 s.f.)

  b t = 4.3 s (2 s.f.)

Exercise 4F

1 **a** d =3000 m (1 s.f.)

  b a = −24.7 m s^{-2} (3 s.f.)

  c v_f = 274 m s^{-1} (3 s.f.)

2 **a** d = 3.6 m

  b t = 1.7 s

  c t = 0.5 or 1.2 s

3 **a** v_h = 19.92 m s^{-1}

  b v_v = 16.71 m s^{-1}

  c d_v = 14 m (2 s.f.)

  d t = 3.4 s (2 s.f.)

  e d_h = 68 m (2 s.f.)

4 **a** v_v =60.62 m s^{-1}

  b d = 190 m (2 s.f.)

  c v_h = 70.0 sin30 = 35 m s^{-1}

5 **a** v_h =9.645 m s^{-1}

  b v_v = 9.645 m s^{-1}

  c d = 20.8 m (3 s.f.)

  d v = 15 m s^{-1} (2 s.f.)

  e t = 0.98 s

6 **a** Trial 1: 0.30 s

  d Distance is proportional to time squared

  f m = 4.91 m s^{-2}

     c = 0 m

     d = 4.91 t^2

     (m) (ms^{-2}) (s^2)

  g a = 9.82 m s^{-2}

Exercise 4G

1 **a** τ = 8.82 Nm

  b τ = 8.82 Nm

  c F = 220 N (2 s.f.)

3 τ = 1216 Nm

4 **b** F_B = 882 N and F_A = 588 N

5 **b** m = 6.0 kg

     F_y = 980 N

6 **b** F_L = 49.98 N, F_R = 24.5 N

7 **b** F_v = 22.5 N

  c 0.45 m

Exercise 4H

1 **a** T = 13.7 s

  b a_c = 5.0 m s^{-2}

2 **a** f = 2.4 Hz

  b T = 0.42 s

  c d = 6.91 m

  d v = 16.46 m s^{-1}

     F_c =68.9 N

3 **b** v = 24 m s^{-1}

  d F = 476 N

  e The force will be a quarter of its previous value.

Exercise 4I

1 **a** k = 440 Nm^{-1} (2 s.f.)

  b E_p = 5.6 J (2 s.f.)

2 **a** x = 0.04 m

  b L_o = 0.78 m

  c L_n = 0.79 m

  d Extension to a third. A third of the energy is stored.

3 **a** k = 9800 Nm^{-1}

  c E_p = 9.9 J

  d Extension to double. Energy stored in the strings will double.

4 **c** Force is proportional to deflection.

  d m = 500 kNm^{-1}

     c = 0 N

  e F = 5 x 10^5 x

  f k = 5 x 10^5 Nm^{-1}

  g Area represents the energy stored in the model eardrum

Exercise 4J

1 **a** F_n = 24.25 N

     F_v = 14.0 N

  b W = 780 J (2 s.f.)

  c P = 16 W

2 **a** F_{net} = 0N

  b $F_{\parallel}$ = 8600sin19 = 2800 N

     F_{boys} = 2870 N up the slope

  c P = 1435 W

3 **a** E_k – 81 J

  b $E_{friction}$ = 81 J

  c F = 1350 N

4 **a** a = 5.83 m s^{-2}

  b F = 8925 N

ISBN: 9780170195997

c   P = 124,950 W

5   **d**   Energy is proportional to compression squared.

    f $m = 1125$ J m^{-2}

       $c = 0$ m

    g $E = 1125\ x^2$

    h $k = 2250$ J m^{-2}

CHAPTER 5

Exercise 5A

1   **a**   Sphere of positive charge with electrons embedded it in it

    b   Described in terms of the different number of electrons.

    c   **i**   how the electrons were arranged;

       ii   why different elements give out different colours when heated.

       iii   ionising radiation.

2   **a**   The positive charge is concentrated in a dense positively charged nucleus which contains most of the mass of the atom. The nucleus was surrounded by orbiting electrons.

    b   Both atoms contain electrons; contain a positively charged mass; are electrically neutral; contain subatomic particles.

    c   In Rutherford's model the electrons orbit the positive mass which is concentrated in the nucleus. In Thomson's model the electrons are embedded in a positive mass which is spread out across the whole atom.

    d   Rutherford's model did not contain neutrons in the nucleus; did not have the electrons moving in stable orbits.

3   **b**   Zinc sulfide fluoresces.

    c   Alpha particles are highly ionising and would quickly ionise the air.

    d   Geiger and Marsden made three observations:

       i   The majority of alpha particles passed straight through;

       ii   Some were scattered through large angles, and;

       iii   A very small number (1 in 8000) were deflected by more than 90° - they bounced back!

    e   Given that the foil was several hundred atoms thick the positive charge must have been significantly smaller than the atom. The alpha particles that were scattered through large angles must have encountered a concentrated positive charge. The alpha particles that were scattered backwards must have encountered something with a much larger mass and concentrated positive charge.

4   **a**   An atom will become ionised when it gains or loses an electron.

    b   All the alpha particles would pass through the atoms and experience only small changes in direction.

    c   The majority of the alpha particles passed straight through with little or no deviation.

    d   Some alpha particles were scattered through very large angles, and 1 in 8000 were scattered backwards off the foil.

    e   Rutherford proposed a new model in which the positive charge was concentrated in a dense positively charged nucleus which contained most of the mass of the atom. The nucleus was surrounded by orbiting electrons, held there by the electrostatic force of attraction between the negatively charged electrons and the positively charged nucleus.

Exercise 5B

1   **a**   A = 185 - 181 = 4; Z = 82 - 80 = 2; X = alpha particle or helium ion.

    b   Conservation of proton (atomic) number and nucleon (mass) number.

    c   A nuclide of an element has a specific number of protons AND neutrons. There are a number of reasons why a nuclide might be unstable:

       i   Too few neutrons.

       ii   Too many neutrons.

       iii   Too much energy. If the nucleus has too much

potential and kinetic energy. The emission of radiation will remove energy from the unstable nuclide and the daughter nuclide may have a more stable proton-neutron ratio.

    d   m = 0.21 kg

    e   The number of emissions each second will decrease at the same rate.

2   **a**   A = 4+14-17 = 1; Z = 2 + 7 – 8 = 1; X = proton or hydrogen ion.

    b   8 protons and 9 neutrons.

    c   In a neutral oxygen-17 atom there will be 8 electrons to balance the number of protons. Neutrons have no charge.

3   **a**   A neutron: $_{0}^{1}n$

    b   Conservation of proton (atomic) number and nucleon (mass) number.

    c $_{28}^{60}Ni$

    d   Cobalt-60 is used to sterilise medical equipment because it emits intense gamma radiation which kills all bacteria leaving the equipment sterile and safe to use.

    e   The half-life of a radioactive nuclide is the time taken for the number of undecayed nuclei to fall to half its original number and is independent of the sample size.

    f   t = 18 years (2 s.f.)

    g   The decrease in mass of cobalt-60 results in an increase in the mass of nickel-60 so the mass of the canister and the contents will still be 300 g after 2 half-lives.

    h $_{0}^{1}n \rightarrow\ _{1}^{1}p +\ _{-1}^{0}e$

4   **a**   Paper causes a significant decrease in the amount of radiation that reaches the Geiger counter. This suggests that the source is a strong alpha emitter, as alpha particles are highly ionising so cannot penetrate paper. Thick lead further reduces the count rate which means the source also emits gamma radiation which is much more penetrating than alpha or beta radiation, due to its extremely weak ionising ability.

    b $_{90}^{230}Th \rightarrow\ _{88}^{226}Ra +\ _{2}^{4}\alpha +\ _{0}^{0}\gamma$

    c   Average half-life: 1600 years

5   **a**   A = 222 – 4 = 218; Z = 86 – 2 = 84; X = polonium.

    b $M_0 = 0.03 \times 25 = 0.96$ g

    c $M_{daughter} = 0.96 – 0.03 = 0.93$ g

    d   Gamma radiation is very weakly ionising and so most of it will pass straight through our bodies without interacting with any of the atoms.

6   Alpha radiation is highly ionising so would cause significant damage to our body cells and would not be able to penetrate our bodies to reach a detector. Beta radiation is weakly ionising, but would ionise cells on its journey through the body, causing significant damage and not reaching the detector to form an image. Gamma radiation has an extremely weak ionising ability so is unlikely to affect any cells on its journey through the body and can be detected outside the body. It is essential that patients are exposed to a minimum amount of any radiation, so an ideal tracer should quickly decay to reduce the exposure time. It is also important that the tracer is not stored by the body for long periods of time, as this would also increase the exposure to the radiation.

7   **a**   A = 131 - 0 = 131; Z = 53 – (- 1) = 54; X = xenon.

    b $M = \frac{2.0}{2^{17}} = 1.5 \times 10^{-5}$ g

    c   Badge 1 has been exposed to alpha radiation.

       Badge 2 has been exposed to alpha and beta radiation.

    d   The open window allows light to strike the film and acts as a control to ensure that the film is reacting to radiation. If the film is operating normally, radiation (in the form of alpha, beta, gamma or light) will cause a chemical reaction to take place gradually turning the film black. The greater the amount of radiation striking the film the darker it goes. The thin plastic window will allow alpha, beta and gamma radiation through but not light. The aluminium window will stop alpha radiation but allow beta and gamma radiation as they are less ionising so more penetrating. The copper disk is much denser so can stop alpha and beta radiation but will allow gamma radiation to pass through as gamma is a very weak ioniser so can penetrate through much denser materials.

  e To reduce the risk to people who are regularly exposed to radiation the following safety measures can be taken:
  i Reduce exposure time,
  ii Wear protective clothing,
  iii Place shielding between the radioactive source and themselves,
  iv Reduce close contact by using remote controlled tools.

  f Ionising radiation interacts with electrons causing ions or free radicals to form inside tissue cells which result in unwanted chemical reactions which can damage or kill the cell. They may also cause uncontrolled cell division and the subsequent growth of a tumour. Low level doses of radiation have little effect on the human body, but doses can be cumulative and it is important that people are aware of their on-going exposure.

8 a The main sources of background radiation that could affect the experiment are due to radioactive gases in the air, radiation from space (cosmic rays), natural sources and building materials. The background radiation will increase the amount of radiation detected during the test and result in a zero error in the measurements of the new and old kauri samples.

  b 15,000 years (2 s.f.)

  c X-rays, neutrons, alpha particles, beta particles and gamma rays are all types of ionising radiation. When ionising radiation encounters an atom it may cause an electron to be removed from the atom, forming a positively charged ion and a free electron. This involves the radiation losing energy and the electron and atom gaining energy.

  d Magnetic fields can deflect both alpha and beta radiation because they are charged particles, but it cannot deflect gamma radiation as it has no charge. As alpha and beta particles have opposite charges they will be deflected in opposite directions. By moving the detector to either side to find the deflected direction, and using Fleming's left-hand rule (or slap rule), the type of charge can be determined.

  e The type of radiation could be determined using different absorbers. Alpha particles have a strong ionising effect so are easily absorbed by a thin sheet of paper. Beta particles have only a weak ionising effect so are more penetrating and require a thick sheet of aluminium to stop them. Gamma radiation has a very weak ionising effect so is very penetrating; consequently it requires thick sheets of lead to decrease its intensity.

9 a A = 241 − 4 = 237; Z = 95 − 2 = 93; X = Neptunium.
  b Alpha particles have a strong ionising effect so will only travel a few centimetres inside the smoke detector before being completely absorbed by the air.
  c t = 50 years.

Exercise 5C

1 a $^{1}_{0}n + ^{235}_{92}U \rightarrow ^{236}_{92}U \rightarrow ^{141}_{56}Ba + ^{93}_{36}Kr + 2^{1}_{0}n + $ energy

  b The strong force provides a very strong but short range attractive force to hold the nucleons together. When the nucleus becomes elongated the strong force is unable to overcome the the longer range electrostatic force which is causing repulsion between the positively charged protons in the nucleus. The result is that the nucleus splits into two smaller nuclei called fission fragments which are more stable.

  c The fission of Uranium-235 results in the emission of two or three fast moving high energy neutrons. However uranium-235 requires slow moving neutrons to undergo

fission, so the reactor vessel is filled with a moderator made of heavy water. The heavy water surrounds the fuel rods and when a fast moving neutron collides with a deuterium nucleus, some of its kinetic energy is transferred to the deuterium causing the neutron to slow down. These slow moving neutrons may result in further fissions in neighbouring uranium-235 nuclides, producing more energy and emitting even more neutrons. If the mass of uranium is too small then most of the neutrons will escape from the uranium before causing more fission to occur. But if there is a critical mass of uranium then each neutron can cause a further fission resulting in a chain reaction. The number of fissions taking place each second can be controlled using boron or cadmium which are effective absorbers of neutrons. In this way only one neutron is available to cause another fission so energy is released at a steady rate.

  d E = 28.8 × 10⁻¹² J
  e 27.1 x 10⁻⁶ kg

2 a Neutrons released as the result of a nuclear fission process are travelling very fast and must slow down before they can take part in further fission reactions. A thermal neutron is a slow moving neutron that is travelling at a speed associated with gas particles moving at normal temperatures.
  b 9.768 x 10²⁷ nuclei
  c E = 27.8 × 10⁻¹² J
  d 5.00 x 10⁶ s
  e This assumes that the power station is able to react all the U-235 in the fuel rods; that no energy is lost to the surroundings as waste heat; that no work is done against friction inside the generator.

3 a $2^{\ 0}_{-1}\beta$ or $2^{\ 0}_{-1}e$
  b Plutonium and silver are examples of elements. Each element has a unique number of protons (atomic number). The reaction above has two different nuclides of the element silver. Each nuclide of an element has the same number of protons but a unique number of neutrons (unique mass number). Collectively the nuclides are referred to as isotopes of silver.
  c E = 30.5 × 10⁻¹² J
  d Conservation of mass-energy.
  e If the mass of plutonium is too small then most of the neutrons will escape from the plutonium before causing more fission to occur. But if there is a critical mass of plutonium then each neutron can cause a further fission resulting in a chain reaction.

4 a Protium, deuterium and tritium are three different hydrogen nuclides. Each nuclide of an element has the same number of protons but a unique number of neutrons (unique mass number). Collectively the nuclides are referred to as isotopes of hydrogen.
  b $^{2}_{1}H + ^{3}_{1}H \rightarrow ^{4}_{2}He + ^{1}_{0}n$
  c The long range electrostatic force of repulsion between the positive nuclei will act to keep the nuclei apart, so a large amount of energy is required to do work against the electrostatic force and push the two nuclei close enough so that the short range strong force is able to bind them together. The strong force is about 100 times stronger than the electrostatic force but acts over a much shorter range.
  d E = 2.82 × 10⁻¹² J
  e 5.92 x 10⁻⁶ kg
  f Both reactions release the same amount of energy but the fusion reaction requires and the fission reaction requires which is nearly 5 × the mass. This demonstrates that fusion reactions release far more energy per kilogram of fuel.
  g 3.89 x 10¹⁶ kg
     5.83 x 10¹⁶ kg

ISBN: 9780170195997

CHAPTER 6

Exercise 6A

1   a $a = -3.2 \times 10^{-14}$ C
2   b $n = 3.0 \times 10^{9}$ electrons
5   a $E = 2 \times 10^{4}$ V m^{-1} (1 s.f.)
    b $V = 5.4$ kV (2 s.f.)
    c $d = 0.15$ m (2 s.f.)
6   a $F = 3 \times 10^{-14}$ N (1 s.f.)
    b   i $F = -8 \times 10^{-15}$ N (1 s.f.)
    c   i $E = 33$ kN C^{-1} (2 s.f.)
       ii   V m^{-1} or N C^{-1}
    d $q = 3.2 \times 10^{-19}$ C
7   a $J = 2 \times 10^{-16}$ J (1 s.f.)
    b   i   opposite direction
       ii $E = 70$ kV m^{-1} (2 s.f.)
    c   i $E = 1.0 \times 10^{-16}$ (2 s.f.)
       iii $d = 3$mm (1 s.f.)
8   b $E = 3.00 \times 10^{6}$ V m^{-1}
    d $n = 6.25 \times 10^{12}$ electrons
    e $E_p = 0.30$ J (2 s.f.)
    g $v = 2.0$ m s^{-1} (2 s.f.)
9   b $\Delta E_p = 4.00 \times 10^{-16}$ J (3 s.f.)
    c $m = 9.05 \times 10^{-31}$ kg (3 s.f.)
    d $V = 500$ V (2 s.f.)
    e $F = -2.00 \times 10^{-15}$ N (3 s.f.)
10   b   electric field is much stronger at the nozzle
    e 1.001×10^{-7} J, 8.336×10^{-13} C, 1.2×10^{5} V
11   a   Trial 1: 4.402 and Trial 2: 3.999 kV m^{-1}
    c   Electric field strength is inversely proportional to distance.
    f $m = 120$ V
       $c = 0$ kV m^{-1}
       $\frac{E}{(V\,m^{-1})} = \frac{120}{d} \frac{(V)}{(m)}$
    g $V = 120$ V

Exercise 6B

1   a $I = 1.8$ A
    b $t = 1.40 \times 10^{3}$ s (3 s.f.)
    c $Q = 75$ C
    d   i 2450 mA h $= \frac{2450 \times 60 \times 60}{1000} = 8820$ A s
         1 A s $= 1$ C so the quantity is a measure of charge.
       ii $n = \frac{8820}{1.6 \times 10^{-19}} = 5.5 \times 10^{22}$ electrons
2   a $V = 240$ V (2 s.f.)
    b $\Delta = 1700$ J (2 s.f.)
    c $Q = 5.4 \times 10^{3}$ C
3   a $R = 48 \, \Omega$ (2 s.f.)
    b $V = 3$ V (1 s.f.)
    c $I = 14$ A (2 s.f.)
4   a $\Delta E = 72$ J (2 s.f.)
    b $V = 2 \times 10^{1}$ V (1 s.f.)
    c $Q = 7.2 \times 10^{4}$ C (2 s.f.)

Exercise 6C

1   a $30.0 \, \Omega$
    b $50 \, \Omega$
    c $32 \, \Omega$
2   a $4.0 \, \Omega$
    b $60.0 \, \Omega$
    c $128 \, \Omega$
3   a $R_p = 4.0 \, \Omega$
       $R_T = 10.0 \, \Omega$
    b $R_p = 10 \, \Omega$
       $R_T = 60 \, \Omega$
    c   i $R_? = 28 \, \Omega$
    d $R_p = 15 \, \Omega$, $R_1 = 45 \, \Omega$
4   a $I = 15$ A

b $R = 0.80 \, \Omega$
d $\Delta E = 43200$ J
e $R = 0.10 \, \Omega$
f $P = 142.2$ W
g $R = 1.3 \, \Omega$
5   a $V = 3.0$ V
    d $R_T = 6.0 \, \Omega$
    e $R = 4.5 \, \Omega$
    f $\Delta E = 9720$ J
6   a $\Delta E = 22$ kJ (2 s.f.)
    b $R = 2.0 \, \Omega$
    c   Three quarters of the power output from a single lamp
    d $6.00 \, \Omega$
    e 0.40 A
7   a   No significant anomalous points
    c   Resistance is inversely proportional to the diameter squared.
    f $m = 0.089 \, \Omega$ m^2
       $c = 0 \, \Omega$
       $\frac{R}{(\Omega)} = \frac{0.089}{d^2} \frac{(\Omega\,m^2)}{(m^2)}$
    g $\rho = 0.47 \, \Omega$ m

Exercise 6D

1   a $F = 0.18$ N (2 s.f.)
    b   i $B = 4 \times 10^{1}$ T (1 s.f.)
       ii   T or N A^{-1} M^{-1}
    c $d = 167$ m
2   a   to the right
    b   to the left
    c   to the left
    d   No force
3   i $F = 1.1$ N (2 s.f.)
    ii   Force acts to the right
4   $I = 0.40$ A
    $B = 2.3$ T
5   d $F = 0.17$ N

Exercise 6E

1   a $V = 575$ m s^{-1}
    b $F = 1.00 \times 10^{-16}$ N
    c $n = 3.0$ electrons
2   a   Downwards
    b   To the left
    c   To the right
    d   No force
3   a   i   Positive
       ii   No charge
       iii   Negative
    d 1.5×10^{7} m s^{-1}
4   b $F = 4.8 \times 10^{-15}$ N
    d   Half the radius

Exercise 6F

1   a $V = 0.49$ V
    b $v = 67$ m s^{-1}
    c $L = 0.69$ m
2   a   Current is down
    b   Current is to the right
    c   No current is induced
    d   No current is induced
3   a   Front (left) negative
    b $V = 0.77$ V, $I = 1.5$ A
4   a $I = 1.75$ A
    c $F = 0.70$ N
    d $W = 3.5$ J

Thomson's model of the atom

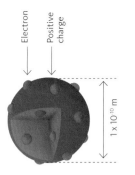

- Electron
- Positive charge

1×10^{-10} m

Rutherford's model of the atom

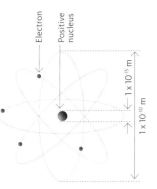

- Electron
- Positive nucleus

1×10^{-15} m

1×10^{-10} m

Rutherford's observations

Majority of alpha particles passed straight through – atom is mostly empty space.

Some alpha particles scattered at large angles – nucleus is positive.

1 in 8000 bounced back – nucleus contains all the mass.

Representing elements

$$^{A}_{Z}X$$

- Element X
- Nucleon number A
- Proton number Z

NELSON
A Cengage Company

New House

ISBN: 9780170195997

L2 Physics
Atomic formula

$$E = mc^2$$

$$P = \frac{\Delta E}{t}$$

Useful equations which are not provided:

$$N = A - Z$$

$$^{A}_{Z}X \longrightarrow \; ^{A-4}_{Z-2}Y \; + \; ^{4}_{2}\alpha \; + \text{ energy}$$

$$^{A}_{Z}X \longrightarrow \; ^{A}_{Z+1}Y \; + \; ^{0}_{-1}\beta \; + \text{ energy}$$

$$^{A}_{Z}X^{*} \longrightarrow \; ^{A}_{Z}X \; + \; ^{0}_{0}\gamma$$

$$N_n = \frac{N_0}{2^n}$$

$$\Delta m = m_f - m_i$$

Constants

$$c = 3.00 \times 10^8 \text{ m s}^{-1}$$

$$q_e = -1.60 \times 10^{-19} \text{ C}$$

$$m_e = 9.11 \times 10^{-31} \text{ kg}$$

$$q_p = 1.60 \times 10^{-19} \text{ C}$$

$$m_p = 1.67 \times 10^{-27} \text{ kg}$$

NELSON
A Cengage Company

New House

ISBN: 9780170195997

L2 Physics
Electricity equations

$$F = Eq$$

$$E = \frac{V}{d}$$

$$\Delta E_p = qEd$$

$$E_k = \tfrac{1}{2}mv^2$$

$$I = \frac{q}{t}$$

$$V = \frac{\Delta E}{Q} \qquad V = IR$$

$$P = IV \qquad P = \frac{\Delta E}{t}$$

$$R_T = R_1 + R_2 + \dots$$

$$\frac{1}{R_T} = \frac{1}{R_1} + \frac{1}{R_2} + \dots$$

$$F = BIL \qquad F = Bqv$$

$$V = BvL$$

Useful equations which are not provided

$$P = I^2R \qquad P = \frac{V^2}{R}$$

$$R = \frac{\rho L}{A} \qquad n = \frac{Q}{q}$$

NELSON
A Cengage Company

New House

ISBN: 9780170195997

L2 Physics
Mechanics equations

$$v = \frac{\Delta d}{\Delta t} \qquad a = \frac{\Delta v}{\Delta t}$$

$$v_f = v_i + at$$

$$d = v_i t + \tfrac{1}{2}at^2$$

$$d = \left(\frac{v_i + v_f}{2}\right)t$$

$$v_f^2 = v_i^2 + 2ad$$

$$a_c = \frac{v^2}{r}$$

$$F = ma \qquad \tau = Fd$$

$$F = -kx \qquad F_c = \frac{mv^2}{r}$$

$$p = mv \qquad \Delta p = F\Delta t$$

$$E_p = \tfrac{1}{2}kx^2$$

$$E_k = \tfrac{1}{2}mv^2$$

$$\Delta E_p = mg\Delta h$$

$$W = Fd \qquad P = \frac{W}{t}$$

$$C = 2\pi r$$

ISBN: 9780170195997

L2 Physics
Waves equations

$$\frac{1}{f} = \frac{1}{d_o} + \frac{1}{d_i}$$

$$m = \frac{d_i}{d_o} = \frac{h_i}{h_o}$$

$$n_1 \sin\theta_1 = n_2 \sin\theta_2$$

$$\frac{n_1}{n_2} = \frac{v_2}{v_1} = \frac{\lambda_2}{\lambda_1}$$

$$v = f\lambda \qquad v = \frac{d}{t}$$

$$f = \frac{1}{T}$$

Constants

Speed of light,

$$c = 3.00 \times 10^8 \text{ m s}^{-1}$$

Useful equations which are not provided

$$r = 2f$$

$$\frac{n_1}{n_2} = \frac{v_2}{v_1} = \frac{\lambda_2}{\lambda_1} = \frac{\sin\theta_2}{\sin\theta_1}$$

NELSON
A Cengage Company

New House

ISBN: 9780170195997

Periodic Table

Atomic Number → (shown as 1), Relative Atomic Mass (shown as 1.0)

1	2											13	14	15	16	17	18
1 **H** 1.0																	2 **He** 4.0
3 **Li** 6.9	4 **Be** 9.0											5 **B** 10.8	6 **C** 12.0	7 **N** 14.0	8 **O** 16.0	9 **F** 19.0	10 **Ne** 20.2
11 **Na** 23.0	12 **Mg** 24.3	3	4	5	6	7	8	9	10	11	12	13 **Al** 27.0	14 **Si** 28.1	15 **P** 31.0	16 **S** 32.1	17 **Cl** 35.5	18 **Ar** 40.0
19 **K** 39.1	20 **Ca** 40.1	21 **Sc** 45.0	22 **Ti** 47.9	23 **V** 50.9	24 **Cr** 52.0	25 **Mn** 54.9	26 **Fe** 55.9	27 **Co** 58.9	28 **Ni** 58.7	29 **Cu** 63.5	30 **Zn** 65.4	31 **Ga** 69.7	32 **Ge** 72.6	33 **As** 74.9	34 **Se** 79.0	35 **Br** 79.9	36 **Kr** 83.8
37 **Rb** 85.5	38 **Sr** 87.6	39 **Y** 88.9	40 **Zr** 91.2	41 **Nb** 92.9	42 **Mo** 95.9	43 **Tc** 98.9	44 **Ru** 101	45 **Rh** 103	46 **Pd** 106	47 **Ag** 108	48 **Cd** 112	49 **In** 115	50 **Sn** 119	51 **Sb** 122	52 **Te** 128	53 **I** 127	54 **Xe** 131
55 **Cs** 133	56 **Ba** 137	71 **Lu** 175	72 **Hf** 179	73 **Ta** 181	74 **W** 184	75 **Re** 186	76 **Os** 190	77 **Ir** 192	78 **Pt** 195	79 **Au** 197	80 **Hg** 201	81 **Tl** 204	82 **Pb** 207	83 **Bi** 209	84 **Po** 210	85 **At** 210	86 **Rn** 222
87 **Fr** 223	88 **Ra** 226	103 **Lr** 262	104 **Rf** 261	105 **Db** 262	106 **Sg** 263	107 **Bh** 264	108 **Hs** 265	109 **Mt** 268	110 **Ds** 271	111 **Rg** 272	112 **Cn** (277)						

Lanthanide Series

57 **La** 139	58 **Ce** 140	59 **Pr** 141	60 **Nd** 144	61 **Pm** 147	62 **Sm** 150	63 **Eu** 152	64 **Gd** 157	65 **Tb** 159	66 **Dy** 163	67 **Ho** 165	68 **Er** 167	69 **Tm** 169	70 **Yb** 173

Actinide Series

89 **Ac** 227	90 **Th** 232	91 **Pa** 231	92 **U** 238	93 **Np** 237	94 **Pu** 239	95 **Am** 241	96 **Cm** 244	97 **Bk** 249	98 **Cf** 251	99 **Es** 252	100 **Fm** 257	101 **Md** 258	102 **No** 259

Constants

$q_e = -1.602 \times 10^{-19}$ C

$q_p = +1.602 \times 10^{-19}$ C

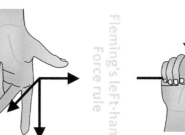

Right-hand grip rule

Fleming's leFt-hand Force rule

Fleming's riGht-hand Generator rule

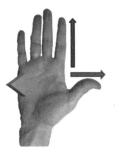

Alternative slap rule

Useful equations which are not provided:

$F_g = mg \qquad W = \Delta E$

$f = \dfrac{1}{T} \qquad T = \dfrac{1}{f}$

$E_p = \frac{1}{2}Fx \qquad d = v_f t - \frac{1}{2}at^2$

Trigonometry

$\sin \theta = \dfrac{\text{opp}}{\text{hyp}}$

$\cos \theta = \dfrac{\text{adj}}{\text{hyp}}$

$\tan \theta = \dfrac{\text{opp}}{\text{adj}}$

Pythagoras

$\text{hyp}^2 = \text{opp}^2 + \text{adj}^2$

Constants

Acceleration of gravity on Earth

$g = 9.8 \, \text{m s}^{-2}$

Ray diagram summary

1 Parallel ray → focal point
2 Focal point → parallel
3 Centre → no change in path

Refraction summary

When the refractive index increases, the speed, angle and wavelength all decrease.

Transverse waves

Displacements in the medium are perpendicular to the direction of travel of the wave.

e.g. surface waves, waves in strings, EM waves, seismic S-waves

Longitudinal waves

Displacements in the medium are parallel to the direction of travel of the wave.

e.g. sound waves, waves in springs, seismic P-waves

Diffraction

Fringe separation increases with:
1 Increasing wavelength
2 Increasing slit-screen distance
3 Decreasing slit separation.